CHEMICAL ENGINEERING METHODS AND TECHNOLOGY

XYLENES

SYNTHESIS, CHARACTERIZATION AND PHYSICOCHEMICAL PROPERTIES

Chemical Engineering Methods and Technology

Additional books in this series can be found on Nova's website under the Series tab.

Additional e-books in this series can be found on Nova's website under the e-book tab.

Environmental Science, Engineering and Technology

Additional books in this series can be found on Nova's website under the Series tab.

Additional e-books in this series can be found on Nova's website under the e-book tab.

CHEMICAL ENGINEERING METHODS AND TECHNOLOGY

XYLENES

SYNTHESIS, CHARACTERIZATION AND PHYSICOCHEMICAL PROPERTIES

MICHAEL OLAWALE DARAMOLA
EDITOR

nova publishers
New York

Library of Congress Cataloging-in-Publication Data

ISBN: 978-1-62808-342-2

Published by Nova Science Publishers, Inc. † New York

CONTENTS

PREFACE

For about three decades, there has been a great evolution in xylenes production technology, with many equipment improvements being instituted in the industry. Typically, these improvements bring economic as well as processing advantages to the producers. Xylene isomers are important chemical intermediates. Para-xylene is almost exclusively used as raw material in the production of terephthalic acid (TPA) and dimethyl terephthalate (DMT), which are reacted with ethyleneglycol to form polyethylene terephthalate (PET), the raw material for polyester resin. Polyester resin is used to manufacture polyester fibres, films and fabricated items (e.g. beverage bottles). According to Tecnon OrbiChem, world para-Xylene demands are expected to rise at an average rate of 7% per year in the period 2008-2013, driven mainly by TPA and PET demand increase in China, other Asian countries and in the Middle East. Asian markets are foreseen as particularly tight, with demands exceeding the supply. Also, application of Xylenes as solvent is found in printing (as component of ink), rubber (common component of rubber) and leather industries (as component of adhesive, paints and varnishes). In addition, laboratory use of xylenes include the use in making dry ice to cool reaction vessels, and as a solvent to remove synthetic immersion oil from microscope objective during light microscopy.

Therefore, this e-book contributes significantly in the area of synthesis, characterization and uses of xylenes. It provides readers an overview of synthesis, applications and treatment of xylene-contaminated media. Greener technique of synthesizing xylene from biomass feedstock is discussed. Some energy-efficient processes, such as membrane-based process for the synthesis of xylene are elucidated. Application of xylenes in the production of nanomaterials, such as carbon nanotube is mentioned and elaborated. The e-book also describes and discusses various techniques for the biodegradation /bioremediation of xylene contaminated media. Various studies documented in this book are thought-provoking and could provide a platform for developing further research efforts towards the development of greener and energy-efficient industrial processes for producing xylenes. Therefore, this book will be of interest to industrialists in the area of xylene production, as well as graduate students and researchers in the industry and in the academia working in the field of chemical engineering, chemistry, and environmental engineering.

In many ways, editing this book has been a privilege and a unique experience. Thanks firstly to our excellent contributors without whose support, this book would not have materialized. It is most fitting that this technological work is published from contributors around the globe and is founded on the spirit of free enquiry coupled with hard work and imagination. It has indeed been a great pleasure to be in touch with all contributors during the last eight (8) months to one (1) year. Thanks also for their patience and understanding. I would be utterly remiss if I did not acknowledge my colleagues who have provided me with the inspiration, motivation and never-ending encouragement by peer-reviewing the chapters from the contributors. In addition, I wish to thank my wife, Omotayo for her unreserved support and patience during the period of editing this book.

Michael O. Daramola, PhD
School of Chemical and Metallurgical Engineering,
Faculty of Engineering & Built Environment,
University of the Witwatersrand,
Private Bag 3, WITS 2050,
Johannesburg, South Africa

In: Xylenes
Editor: Michael Olawale Daramola

ISBN: 978-1-62808-342-2

Chapter 1

XYLENES: PRODUCTION TECHNOLOGIES AND USES

E. F. Aransiola[1,3], M. O. Daramola[2] and T. V. Ojumu[3]

[1]Biochemical and Reactions Engineering, Department of Chemical Engineering, Obafemi Awolowo University, Ile-Ife 220005, Osun State, Nigeria

[2]School of Chemical and Metallurgical Engineering, Faculty of Engineering & the Built Environment, University of the Witwatersrand, Johannesburg, South Africa

[3]Department of Chemical Engineering, Cape Peninsula University of Technology, Cape Town, South Africa

ABSTRACT

In the recent past, the use of Xylene in industrial, domestic, clinical and agriculture is becoming important. It is an aromatic hydrocarbon which consists of three distinct isomers; paraxylene, orthoxylene and metaxylene, and all these isomers have been found useful. The world demand for xylene is increasing at a very fast rate; therefore the increase in its demand has created a challenge that needs to be addressed in the industrial sector. This chapter, therefore presents a short review on the usefulness of xylene in various industrial sectors with the aim of looking into various sources and techniques of producing xylenes to meet the world demand.

Keywords: Xylenes, para-xylene, ortho-xylene, meta-xylene, xylene uses

INTRODUCTION

Xylene is an aromatic hydrocarbon molecule containing a benzene ring with two methyl side chains, formula $C_6H_4(CH_3)_2$. It is also known as xylol, dimethylbenzene, violet 3, methyl toluene, commercial or mixed xylene. It is commonly referred to as xylenes, due to the fact that it is an isomer that can be found in three forms; meta-xylene (*m*-xylene), ortho-xylene (*o*-xylene) and para-xylene (*p*-xylene). The trade names of these isomers are written in

Table 1. Commercial or mixed xylene generally contains about 40–65% *m*-xylene and up to 20% each of *o*-xylene, *p*-xylene, and ethylbenzene as well as small quantities of toluene [1].

Xylene is primarily a synthetic chemical, named in 1851, having been discovered as a constituent of wood tar. Several million tons of these are produced annually [3]. Chemical industries produce xylene from petroleum. It also occurs naturally in petroleum and coal tar and is formed during forest fires. It traditionally comes in a mixed form of all the three isomers, each having two methyl groups attached to a basic benzene hydrocarbon ring. These isomers are differentiated by the position of the methyl groups on the ring. Para-xylene has the methyl groups attached on opposite sides of the ring; ortho-xylene has them next to each other, while meta-xylene has the two methyl groups positioned with one carbon in between them (See Table 1) from chemical identity table.

Xylene represents about 0.5 - 1% of crude oil, depending on the source, it is hence found in small amounts in gasoline and airplane fuels. Xylene is mainly produced as part of the BTX aromatics (benzene, toluene and xylenes) extracted from the product of catalytic reforming known as "reformate". The mixture is a slightly greasy, colourless liquid commonly encountered as a solvent. In 2011, a global consortium began construction of one of the world's largest xylene plants in Singapore [4].

The three xylenes and the isomeric ethyl benzene exhibit similar properties due to their similar structure. An importance consideration is given to the distillation characteristics of the C_8 aromatic compounds as it helps in the separation of one from the other. m-xylene and o-xylene are more readily separated from each other because of the 5°C difference in boiling point. p-xylene is easily separated from other C_8 aromatic compounds due to the difference in their freezing points. The critical compression ratios are 14.2, 13.6, and 9.6 for p-xylene, m-xylene and o-xylene respectively. The research octane values are 113, 116.4, 117.5 and 107.4 for Ethyl benzene, p-xylene, m-xylene and o-xylene respectively [5]. The physical properties differ slightly from isomer to isomer, taking for example the boiling point, melting point and the density of the three isomers. Most reactions involving the position of the alkyl substituents include isomerization, disproportionation and dealkylation. According to [5], acids catalyze the interconversion of the three-xylene isomers. Xylenes isomerize to near equilibrium levels in a hydrogen fluoride – boron trifluoride system with low boron trifluoride concentrations. Isomerization at lower temperatures produces more p-xylene and o-xylene. The physical and chemical properties of these compounds are summarized in Table 2.

Table 1. Chemical Identity of Mixed Xylene, *m*-Xylene, *o*-Xylene and *p*-Xylene (Adapted from [2])

Characteristics	*m*-Xylene	*o*-Xylene	*p*-Xylene
Synonyms/trade names	1,3-Dimethyl benzene; m-dimethylbenzene; m-methyltoluene; 1,3-xylene; m-xylol; meta-xylene	1,2-Dimethyl benzene; o-dimethylbenzene; o-methyltoluene; 1,2-xylene; o-xylol; ortho-xylene	1,4-Dimethyl benzene; p-dimethylbenzene; p-methyltoluene; 1,4-xylene; p-xylol; para-xylene
Chemical formula	C_8H_{10}	C_8H_{10}	C_8H_{10}

Table 2. Physical and Chemical Properties of Mixed Xylene, *m*-Xylene, *o*-Xylene and *p*-Xylene (Adapted from [2])

Characteristics	Mixed Xylene	*m*-Xylene	*o*-Xylene	*p*-Xylene
Molecular weight	106.16	106.16	106.16	106.16
Color	Clear	Colorless	Colorless	Colorless
Physical State	Liquid	Liquid	Liquid	Liquid
Melting Point	No data	−47.8°C	−25.2°C	13.2°C
Boiling Point	137-140°C	139.1°C	144.5°C	138.4°C
Density at 20°C	0.864 g/cm^3	0.864 g/cm^3	0.880 g/cm^3	0.861 g/cm^3
Odor	Sweet	Sweet	Sweet	Sweet
Odor threshold: Water Air	No data 0.0045 mg/L (1.0 ppm)	No data 0.05 ppm	No data 0.05 ppm	No data 0.05 ppm
Solubility: Water at 25°C Organic solvent (s)	106 mg/L Miscible with alcohol and ether	161 mg/L Miscible with alcohol, ether and other solvents	178 mg/L Miscible with alcohol and ether	162 mg/L Soluble in alcohol, ether and other organic solvents
Partition Coefficients: Log K_{ow} Log K_{oc}	No data No data	3.2 2.22	3.12 2.11	3.15 2.31
Vapour Pressure	6.72 mmHg at 21°C	8.29 mmHg at 25°C	6.61 mmHg at 25°C	8.84 mmHg at 25°C
Henry's law constant	No data	7.18×10^{-3} atm m^3/mol	5.18×10^{-3} atm m^3/mol	6.90×10^{-3} atm m^3/mol
Autoignition temperature	464°C	527°C	463°C	528°C
Flash point	29°C	27°C	32°C	27°C
Flammability limits	No data	1.1-7.0%	1.0-7.0%	1.1-7.0%

Current World Production and Demand for Xylenes

The worldwide growth rates for xylenes were 7% and 8.8% in the 1990 to 1995 and 1995 to 2000 periods, respectively [6]. The differential increase for the world production and demand for xylenes between year 2003 and 2009 were 4.7 and 10.1 million tons respectively. The details of this growth rate in Global Total (GT), Asian Total (AT), Western Europe (WE), North America (NA) and Middle East (ME) are shown in Figures 1 and 2. The Asian continent indicated includes South Korea (SK), Taiwan, China, ASEAN (Association of Southeast Asian Nations), India and Japan. In the year 2010, world production of xylenes, both pure and mixed isomers, made up around 44m tonnes [7]. The xylene consumption increased by 10.6% from 2009 and fixed approximately on a level with the production.

Experts expect the global demand to grow by 4.4% per year from 2011 to 2015 and by 3.2% from 2015 to 2020. The capacity utilisation however is predicted to reduce from 2011 to 2014 but begin an upward tendency afterwards. According to the report by the Process Economics Program compiled in November, 2006 by Marcos Cesar, global demand for xylene isomers has been steadily increasing, with p-xylene leading the way. The report records that p-Xylene is a key raw material for the large polyester fiber industry, which is experiencing global growth rates near 6% per year. Future polyester growth will continue to be driven by the Asia region and, more specifically, by the Chinese market (see Figure 2). After a period of severe over capacity during 1999-2002, global operating rates for p-xylene were above 90% in 2005. Also according to Thyssenkrupp Uhde, Germany [8], the increase in demand of xylene is determined by p-xylene which is the biggest isomer in terms of quantity, and that the growth in consumption of this isomer is expected to well exceed 5% per annum in the near future.

EXISTING INDUSTRY TECHNIQUES TO PRODUCE XYLENES

The existing techniques for the mixed xylene production are discussed in this section. An understanding of the production of the mixed xylene forms the basis of each of the isomers. Mixed xylenes occur naturally in petroleum and coal tar and mostly being derived from petroleum fractions as part of the BTX aromatics. However, the concentration of the BTX aromatics in petroleum rarely exceeds one percent [10]. Crude oil by processing can be converted to BTX streams. Several petroleum fractions are used in aromatic conversion processing. According to Planning and Standards [10] and Hancock [11], the fraction most important to the xylene production process is "straight-run light naphtha" which includes all of the crude oil components heavier than pentanes and up to a final boiling point between 105°C and 170°C (221° to 338°F). It is from this stream that the majority of xylene is produced by catalytic reforming via hydrotreating.

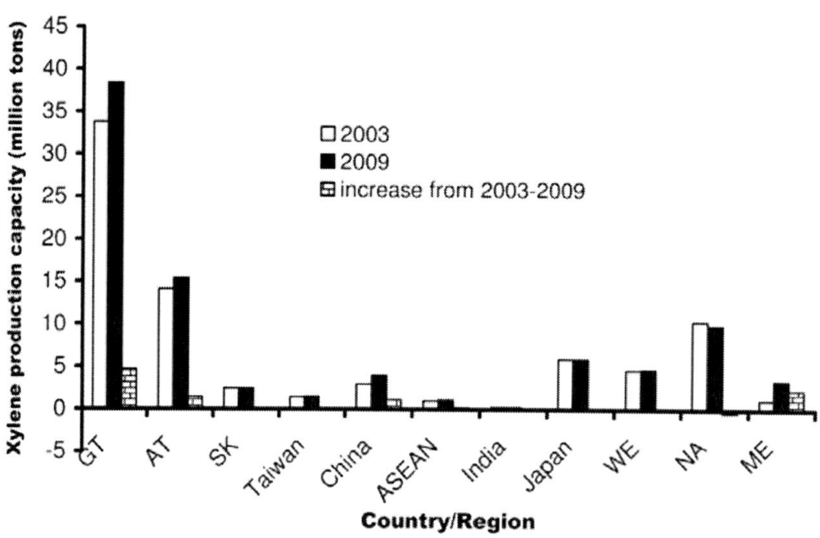

Figure 1. Global Xylene Production Capacity between Year 2003 and 2009 (Adapted from [9]).

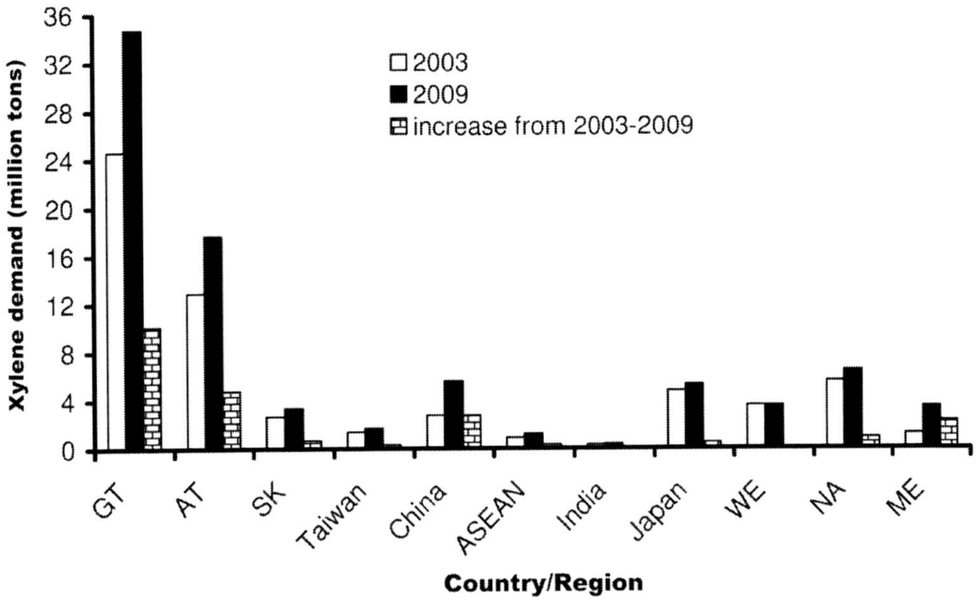

Figure 2. Global Xylene Demand between Year 2003 and 2009 (Adapted from [9]).

The naphtha that results from the pyrolysis or "steam cracking" (*e.g.*, hydrocracking) of heavier distillate fractions is also used as a feedstock in xylene production. It is the secondary reactions of naphtha cracking that produce considerable amounts of "pyrolysis gasoline" rich in aromatics. The additional method for xylene production is the separation from coal tars that is the coal-derived coke-oven light oil and disproportionation or transalkylation of toluene.

Hydrotreating

Hydrotreating, formally known as hydrodesulfurization (HDS), is a chemical process used in removing sulphur from natural gas and refined petroleum streams. This process becomes very important if pure aromatics are to be obtained. Removal of sulphur is required to eliminate a major poison in subsequent processing. It is also a means of improving the quality of the petroleum fractions, and to meet environmental legislation. This is done by subjecting them to mild or severe conditions of hydrogen pressure in the presence of a catalyst. Commercially used catalysts for reformer feedstock hydrotreatment belong to the broad class of molybdenum sulfide catalysts, which are applied for the hydroprocessing of a variety of feedstocks, including middle distillates and gas oils. The major requirements to be met by naphtha hydrotreatment catalysts are high activity and selectivity for hydrogenolysis of CS bonds, with low selectivity for cracking of C-C bonds. In addition, they should exhibit reduced activity for hydrogenation of aromatic compounds in order to maintain octane number and reduce the hydrogen consumption. The catalyst should also possess long-term stability and low sensitivity to poisons [12].

It is necessary for the feeds; pyrolysis gasolines (from secondary hydrogenation), straight-run light naphthas (*e.g.*, catalytic reformer feeds) and coke-oven light oil to undergo hydrotreating prior to subsequent processing and xylene recovery, so as to be able to separate the non aromatics from the aromatics by extractive distillation. In this process, the feedstock combined with the recycled gas and make up hydrogen gas are pressurized and heated to the set pressure and temperature respectively. These are then introduced into a fixed-bed reactor containing catalyst where the hydrogenation reaction takes place. After leaving this reactor, the stream is cooled and moved to a separator vessel where recycle or net hydrogen is removed, then to a stripper. The stripped, hydrotreated fraction will either be used for the catalytic reformation or secondary hydrogenation [10]. Catalytic hydrogenation is become an appropriate technology for eliminating impurities such as diolefins, olefins, sulphur, nitrogen and oxygen components.

Catalytic Reforming of Naphtha

Catalytic naphtha reforming is a very important process for producing high octane gasoline, aromatic feedstock and hydrogen in petroleum-refining and petrochemical industries [13]. Ninety five percent of xylene produced is from catalytic naphtha reforming process. Catalytic reforming involves the catalytic dehydrogenation of straight-run light naphtha in the presence of hydrogen (which reduces coke formation) to yield a mixture of aromatic hydrocarbons (*e.g.*, benzene, toluene, and the xylenes) [10,11,14]. According to Planning and Standards [10], the hydrotreated naphtha is combined with recycled hydrogen, preheated and heated to the reaction temperature in a fired heater. This is then transferred to a series of catalyst containing reactors. At this stage, a series of three or four reactors with inter-stage reheat furnaces may be necessary to achieve the required conversion, due to the fact that the reaction is endothermic. The reactors normally contain increasing amounts of catalyst in each stage [11]. The effluent from the last reactor is cooled and transferred to a receiving unit (*e.g.*, the flash drum) where the hydrogen is separated from the liquid reformate. The hydrogen gases are compressed and recycled to the reactors while the reformate is moved to a stabilizer fractionator. The fractionator removes C4 and lighter hydrocarbons to produce a stabilized reformate. The stabilized reformate is used as a feedstock in the xylene recovery process. The process flow diagram of this process is shown in Figure 3. Most of the facilities that produce xylene by catalytic reforming have proprietary processes. Table 3 lists the process licensor and the process name. The primary differences between these processes involve solving reforming process problems such as catalyst regeneration. The processes also differ in the methods used to extract aromatics depending on the type and purity of the product desired. According to Antos and Aitani [12], the catalytic naphtha reforming process can be classified into three types; Semiregenerative, Cyclic (fully regenerative) and Continuous regenerative (moving bed). This classification is based on the frequency and mode of regeneration. The semiregenerative requires unit shutdown for catalyst regeneration, whereas the cyclic process utilizes a swing reactor for regeneration in addition to regular inprocess reactors. The continuous process permits catalyst replacement during normal operation. Table 4 presents a regional distribution of catalytic reforming capacity by process design.

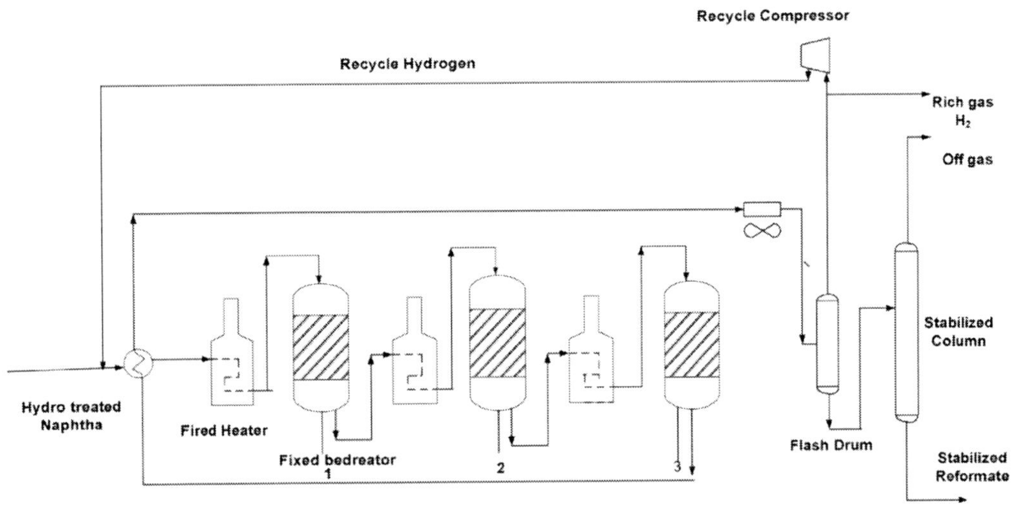

Figure 3. Process Flow Diagram for Catalytic Naphtha (Adapted from [10].

Table 3. Catalytic Reforming Process

Licensor	Name of Process
Chevron Research Co.	Rheniforming
Engelhard	Magnaforming
Exxon Research Engineering	Powerforming
Houdry Division, Air Products	Houdriforming
Institut Francais du Petrole	Aromizing Catalytic Reforming
Standard Oil Co. (Indiana)	Ultraforming
UOP Process Division	Platforming

Source: [10].

Table 4. Regional Distribution of Catalytic Naphtha Reforming by Capacity

Region	Crude Capacity (1000b/d)	Reforming Capacity (1000b/d)	Reforming as % of Crude Capacity
N. America	20,030	4075	20.3
W. Europe	14,505	2135	14.7
Asia – Pacific	20,185	2000	10.0
E. Europe	10,680	1430	13.4
Middle East	6075	570	9.4
S. America	6490	400	6.1
Africa	3200	390	12.1
Total	**81,165**	**11,000**	**13.6**

Source: [12].

Secondary Hydrogenation of Pyrolysis of Gasoline

This technology is used for hydrogenation of the pyrolysis gasoline from naphtha/gas oil steam cracker. Pyrolysis gasoline is an important by-product of ethylene and propylene manufacture produced in the steam cracking unit. This accounts for four percent of domestic xylene production capacity feedstock materials (See Table 5). The yield of pyrolysis gasoline is 0.5~0.8 times of that of ethylene in the unit. The content of aromatics in pyrolysis gasoline is about 60~85% which forms an important aromatics source [15].

The pyrolysis gasoline contains reactive compounds (*e.g.*, diolefins and styrenes) which will polymerize if subjected to reactor conditions severe enough to saturate olefins and remove sulfur compounds. Due to this fact, it must undergo an initial hydrogenation step described in Section 2.1.1 to reduce the reactives to olefins prior to storage or further processing. The resulting product can be used as a high octane gasoline blending component or treated further for aromatic (*e.g.*, benzene, toluene, and xylenes) extraction [10].

The hydrogenation of pyrolysis gasoline is a two stage hydrogenation process by which unsaturated hydrocarbons in pyrolysis gasoline are converted to saturate and the organic sulfur, nitrogen and oxygen in it are removed. Then the effluent is split to different hydrogenated pyrolysis gasoline cuts. The liquid is then stripped of gaseous impurities, such as hydrogen sulfide, and remaining light hydrocarbons before being transferred to xylene recovery units.

Coal-Derived Coke-Oven Light Oil

A very good feedstock for recovering BTX and higher aromatics is coke oven light oil (COLO). In every ton of coal carbonized in coke ovens, about 2 to 3 gallons of a crude light oil is produced that contains 3 to 6 percent mixed xylenes by volume (See Table 5).

This feedstock is first hydrogenated with a special hydro refining process which has a specific heating/evaporation system. A two stage reaction system is used for this purpose. The hydrogenated coke oven light oil is sent to the stabilizer or the hydro refining unit where the off-gas is separated. The stabilized coke oven light oil is sent to a deheptaniser or predistillation unit. The C_7 fraction from this unit is then sent to the extractive distillation unit where benzene and toluene are sepearated from the non aromatics before being separated into high purity benzene and toluene.

The C_{8+} fraction from the bottom of the dehaptanizer is sent to xylene fractionation unit where xylenes and higher aromatics are recovered by distillation [8]. The process flow diagram is shown in Figure 4.

Disproportionation or Transalkylation of Toluene

The process of toluene disproportionation over zeolite ZSM-5, developed by Mobil Co., was reported to be able to effectively upgrade toluene into benzene and xylenes. By transalkylation with C_{9+} aromatics, toluene can be also converted to xylenes. Transalkylation processes are especially attractive to those refineries having BTX extraction units, where excessive C_{9+} aromatics are produced [16]. Less than one percent of recovered xylenes is obtained from toluene disproportionation or transalkylation processes [10]. The toluene disproportionation/transalkylation method of producing xylenes is expensive when compared to the reforming process; however, it has two advantages. One is that no ethylbenzene is formed in the xylenes stream, so isomer isolation is less difficult. Second, no net hydrogen is consumed [10].

Uses and Environmental Impact Assessment of Xylene

Today's market for xylene is growing widely. A lot of industries are using xylenes as feedstocks for the manufacturing of so many day to day products.

Table 5. Typical Composition (wt %) of Aromatics Recovered from Pyrolysis Gasoline, Reformate and Coke Oven Light Oil

Components	Pyrolysis Gasoline	Reformate	Light Reformate	Coke Oven Light Oil
Benzene	30	3	24	65
Toluene	20	13	46	18
Xylene	4	18	< 0.5	6
Ethyl benzene	3	5	< 0.5	2
C_{9+} Aromatics	3	16	0	7
Total Aromatics	60	55	70	98
Naphthenes	High	Low	Low	High
Olefins	High	High	Low	High
Paraffins	Low	High	High	Low
Sulphur	Up to 1000 ppm wt.	< 1 ppm wt.	Low	Up to 1 wt. %

Source: [7].

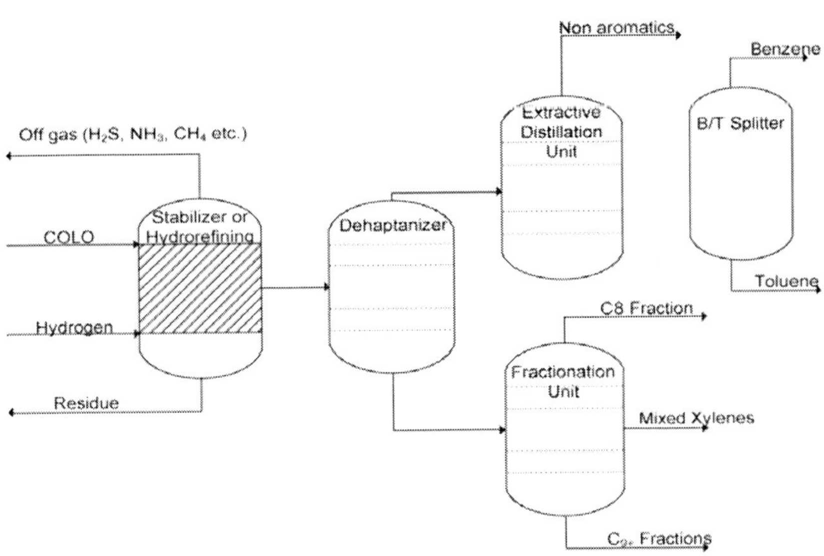

Figure 4. Process Flow Diagram for Aromatics derived from Coke Oven Light Oil (COLO) (Adapted from [8]).

Early Use in the Oil Field

Xylenes had once played a crucial role in the discovery and drilling of oil in the oil fields [17]. Mud drilling had to be performed to test oil concentrations prior to drilling an oil well in the earth. Xylenes were used as solvents to break up and dissolve the mud to allow testing for oil deposits. In recent times, propylene ether is the main solvent for this process.

Xylenes as Solvents in Paint, Ink and Other Industries

Xylenes are one of the important chemicals used as solvents [10, 18]. They are widely used as thinners and solvents in paints, varnishes, adhesives and inks. Xylene is only one of the vehicle solvents used by paint manufacturers. Paints and coatings account for about 65-70 percent of mixed xylenes consumption as solvents. Like toluene, the use of xylene in paints and coatings has been increasing since 1987, largely due to increasing consumption in short-oil and medium oil-length alkyds.

Xylenes mixture can be used to thin lacquers (a black resinous substance used a natural varnish) when slower drying is desired. They are also used as solvents (liquids that can dissolve other substances) in the printing, rubber, and leather industries, as well as in pharmaceuticals, perfumes, fabricated items and pesticide formulations.

Uses of Xylene in the Plastic Industry

All the isomers of xylenes are very useful in the plastic industry. Para-xylene is a key intermediate in the synthesis of purified tetraphthalic acid (PTA) and dimethyl terephthalate (DMT), both of which are used in the production of industrial plastics and polyesters [19]. PTA in particular is used in the production of polyethylene terephthalate (PET) bottle resins [18]. Ortho-xylene is the dominant feedstock used in phthalic anhydride (PA) production [20]. Phthalic anhydride is produced by the oxidation of o-xylene, naphthalene, or mixtures of both feedstocks, in the presence of catalysts. PA is used in the manufacture of a variety of products including plasticizers [6], unsaturated polyester resins, alkyd resins, polyols, phthalocyanine pigments, dyes, perfumes, pharmaceuticals and chemical intermediates. The largest end use of phthalate plasticizers is in compounding flexible polyvinyl chloride (PVC). Other end uses for phthalate plasticizers include some nitrocellulose lacquers and some adhesives. The unsaturated polyester resins are used to produce a number of fabricated fiberglass-reinforced plastics including construction materials, boats and molded automobile body panels (10).

Alkyd resins are derived from phthalic anhydride and used to make surface coatings for machinery, furniture and other materials. However, due to high volatility, this application is being replaced by other coating technologies. The major chemical use of *meta*-xylene is in the manufacture of isophthalic acid, which is used as a copolymerizing monomer to alter the properties of polyethylene terephthalate (PET), making PET more suitable for the manufacture of soft drinks bottles.

Uses of Xylene in the Laboratory

Xylene is used in histology to clean tissue for the preparation of paraffin wax [21]. It is also used to prepare very thin slice of tissues for microscopic examination by making them hydrophobic so as to be able to use a coverslip. It is used in the laboratory to make baths with dry ice to cool reaction vessels, and as a solvent to remove synthetic immersion oil from the microscope objective in light microscopy [21].

Uses of Xylene in the Petroleum Industry

Xylene is used as a feedstock in the production of petrol. It is also found in small proportions in jet fuel. More than 90 percent of mixed xylene isomers are used for petrol blending. This is usually added to gasoline to raise the octane rating, thereby suppressing engine knock, increasing power, and providing smoother running engines [10].

Environmental and Toxicological Aspects of Xylene

Xylene is being identified as a chemical that requires special care when handling, due to toxicity to humans and animals. It may be emitted when xylene-containing products such as paint, ink and gasoline release small amounts over time. In animals, xylene inhalation acts as a central nervous system depression. Xylene in humans acts as a narcotic in high doses. Amounts of xylene over 18,000 ppm can be fatal [17]. Ingesting xylene may cause liver, kidney and gastrointestinal damage. There is no substantive evidence that xylene is carcinogenic. According to Mcmichael [22], the evidence of carcinogenicity of xylene at high concentrations in experimental animals was reported. Mcmichael also reported that any future epidemiological observations of cancer risks associated with xylene would have to take account of the suspected effects of benzene impurities.

Conclusion

In this chapter, world demand for xylene, techniques of producing xylenes, and uses of xylenes in various industrial sectors have been discussed. Also, toxicological effects of xylenes on humans and animals have been enumerated. The need to increase its production across the globe to meet the world demand, and a way to reduce its toxicological effect are great challenges that need to be tackled by scientists and industrialists.

References

[1] Fishbein, L. *Sci Total Environ*. 43, (1985), 165-183.
[2] http://www.atsdr.cdc.gov *Chemical and Physical information on Xylene*.

[3] Fabri; Jörg; Graeser; Ulrich; Simo, TA; (Eds), Xylenes, *Ullmann's Encyclopedia of Industrial Chemistry*, Wiley-VCH, Weinheim, 2002, 28-433.

[4] Tremblay; Jean-François, *Chem. Eng. News*, 89, (2011), 18–19.

[5] http://www.sbioinformatics.com.

[6] Tsai, TC; Liu, SB; Wang, I. *I. Applied Catalysis A: General*, 181, (1999), 355-398.

[7] http://www.chemmarket.info: Global and CIS Production of Xylene Isomers.

[8] http://www.uhde.eu/fileadmin/documents/brochures/uhde_brochures_pdf_en_16.

[9] http://www.meti.go.jp/english/information/downloadfiles/c0517PetroProe.

[10] Planning, USEPAOoAQ. Standards. Locating and Estimating Air Emissions from Sources of Xylene, *Office of Air Quality Planning and Standards*. 1994.

[11] Hancock, EG; ed. Toluene, the Xylenes and Their Industrial Derivatives. *Elsevier*, Scientific Publishing Company. New York, NY. 1982.

[12] Antos, GJ; Aitani, AM. *Catalytic naphtha reforming*. CRC Press. 2004.

[13] Hou, W; Su, H; Hu, Y; Chu, J. Modeling, *Chinese Journal of Chemical Engineering*, 14, (2006), 584-591.

[14] Kirk-Othmer. *Encyclopedia of Chemical Technology*, third ed., Volume 4. John Wiley and Sons. New York, NY. 1978.

[15] http://www.sinopectech.com/news/AtriclePIC/file/Pdf/Petrochemicaltech/Pyrolysis%20 Gasoline%20Hydrogenation%20Technology_ENG.

[16] Wang; Tsai, TC; Huang, ST. *Industrial & engineering chemistry research*, 29, (1990), 2005-2012.

[17] Ramos, N. *Uses for Xylene in the Oilfield*. Available at http://www.ehow.com/ info_8707600_uses-xylene-oilfield.html

[18] http://www.shell.com/chemicals/products-services/ourproducts/aromatics/xylenes/ product-overview.html

[19] Dursch, T; Khalil, R; Khine, A; Mutahi, F. Toluene Methylation To Para-Xylene-Senior Design Reports, (CBE). Department of Chemical & Biomolecular Engineering, University of Pennsylvania. Available at http://repository.upenn.edu/cbe sdr/7 2009.

[20] Santos, KAO; Dantas Neto, AA; Moura, MCPA; Castro Dantas, TN. Brazilian journal of petroleum and gas **5**(2011), 255-268.

[21] http://www.ehow.com/about 5544817 uses-xylene.html

[22] Mcmichael, A. IARC scientific publications, 3, (1988).

In: Xylenes
Editor: Michael Olawale Daramola

ISBN: 978-1-62808-342-2

Chapter 2

PRODUCTION OF XYLENES BY NAPHTHA REFORMING

***Maria do Carmo Rangel*[1]*, *Carlos Luís Pieck*[2], *Luciene Santos Carvalho*[3] *and Mario Ricardo Sad*[2]**

[1]GECCAT, Instituto de Química, Universidade Federal da Bahia, Campus Universitário de Ondina, Federação, Salvador, Bahia, Brazil

[2]Instituto de Investigaciones en Catálisis y Petroquímica, Universidad Nacional del Litoral, Santa Fe, Argentina

[3]Grupo de Pesquisa e Inovação em Química, Instituto Federal da Bahia, Campus Camaçari, Lot. Espaço Alpha, Limoeiro, Camaçari, Bahia, Brazil

ABSTRACT

Naphtha reforming is one of the most important routes to produce high value chemicals for petrochemical industry such as benzene, toluene and xylenes. Among them, xylenes are especially desirable due to their use in several applications such as solvent in printing, rubber and leather industries and components of ink, rubber, adhesive and leather industries. They also can be used as cleaning agents for steel, silicon wafers and integrated circuits. In addition, p-xylene is the principal precursor to terephthalic acid and dimethyl terephthalate, both monomers used in the production of polyethylene terephthalate for plastic bottles and polyester clothing. On the other hand, o-xylene is an important precursor to phthalic anhydride.

The xylene contents in the reformate stream largely depend on the composition of feedstock, the type of reformer, the operating conditions and on the catalysts. The catalysts are especially important, since by changing their properties, one can increase the yield of xylenes or even one of their isomers, by keeping the process conditions. The commercial catalysts are chloride alumina-supported platinum, which have metal and acidic sites. Both catalytic functions act cooperatively for most reactions of naphtha reforming, such as skeletal isomerization and dehydrocyclization of n-alkanes and aromatization of alkyl cyclopentanes. Other metals (rhenium, germanium, iridium and others) are often added to the monometallic catalysts generating new systems able to work under different conditions, making it possible to obtain tailored reformates.

In this chapter, we compare the activity, selectivity and stability of Pt/Al_2O_3 catalysts modified by tin, rhenium or germanium prepared by coimpregnation and successive impregnation for the production of xylenes at atmospheric pressure. As most heterogeneous catalysts, the properties of these catalysts largely depend on the preparation method (successive impregnation or coimpregnation). Also, they depend on the kind of the second and/or third metal (rhenium, germanium, tin), the replacement of water by acetone or ethanol in the impregnation step and the order of precursor addition, during impregnation. It was found that it is possible to design a catalyst able to increase the xylenes yield as well as of some of their isomers.

Keywords: Xylenes, aromatics, naphtha reforming

INTRODUCTION

In recent years, the demand for aromatics, especially xylenes (or dimethyl benzenes), is growing fast mainly due to their numerous applications. The mixture of the xylene isomers (called xylenes or xylol) are extensively used as solvents in printing, rubber and leather industries as well as in odontology. It is also a common component of ink, rubber, adhesive and leather industries, as well as in thinning paints and varnishes for the conservation of art objects. Furthermore, it is used for cleaning steel, silicon wafers and integrated circuits. In the petroleum industry, xylene is a component of paraffin solvents used for removing paraffin wax from clogged tubing. However, the most consumption of xylene mixtures is to produce the individual isomers which found several applications in the synthesis of chemicals. Among the isomers, *p*-xylene and *o*-xylene are by far the highest value ones. The first one is the precursor for the manufacture of terephthalic acid and dimethyl terephthalate both monomers used as starting materials for the production of polyethylene terephthalate (PET), used in the manufacture of plastic bottles and polyester clothing. On the other hand, *o*-xylene is mainly used for the synthesis of phthalic anhydride and for manufacturing plasticizers, which are responsible for the flexibility of polyvinyl chloride (PVC). Furthermore, *m*-xylene can be oxidized to isophthalic acid and be used in technologies for obtaining polyurethane and Nomex aramid polymer (from Dupont), as an electrical insulating in thermally aggressive environments [1].

Xylenes can be obtained by several commercial routes, including naphtha reforming, thermal cracking processes and the reactions of toluene, such as disproportionation, transalkylation and the alkylation with methanol [2-4] as shown in Figure 1. These processes usually produce a mixture of xylenes from which the individual isomer is recovered. The desired purity (>99%) involves high costs to separate the three isomers and ethylbenzene with similar boiling points [5]. Currently, the main routes for the xylenes production are naphtha reforming and toluene disproportionation, the first one account for over 75% of xylenes worldwide.

Reaction Paths in Aromatics Formation

The main goal of catalytic reforming processes is to convert naphthenes and paraffins to isoparaffins and aromatics. The naphtha oil typically contains large amounts of non-branched paraffins and relatively small amounts of branched paraffins, naphthenes, olefins and aromatics. During catalytic reforming, the long chain hydrocarbons are reconstructed through reactions of isomerization, hydrogenation, dehydrogenation and dehydrocyclization, among others [6, 7]. These reactions occur on acidic and/or metallic sites, requiring a bifunctional catalyst, which typically consists of chlorided alumina-supported platinum (Pt/Al_2O_3-Cl). Undesirable reactions, such as hydrocracking, hydrogenolysis and coke formation, also occur decreasing the formation of desired products and forming lights gases (methane, ethane, propane and butane), besides consuming hydrogen and deactivating the catalyst [8, 9].

The main reactions that occur in naphtha reforming are dehydrogenation of paraffins and naphthenes, dehydroisomerization of cyclopentanes, isomerization of paraffins, dehydrocyclization of paraffins, hydrocracking, hydrogenolysis and coke formation. Naphthenes are easily and rapidly dehydrogenated to aromatics by the action of the metallic function. However, due to the high amounts of paraffins in naphtha feed, the most important reaction is the paraffin dehydrocyclization that produces aromatics and hydrogen, and the isomerization which produces isomers with high octane number [7].

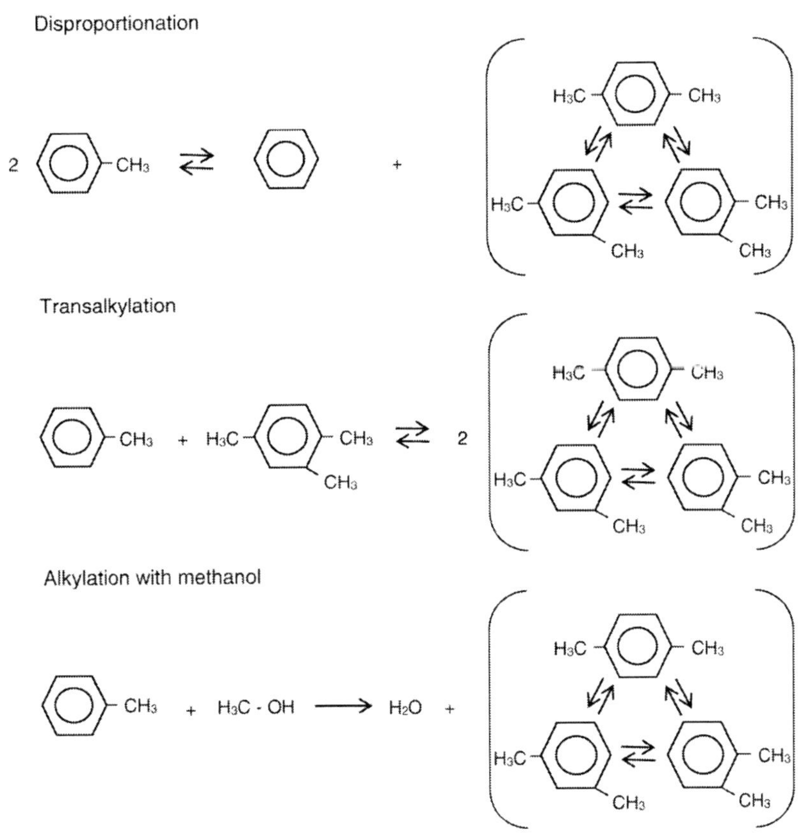

Figure 1. Some reactions of toluene to produce xylenes.

The dehydrocyclization reaction can proceed either by a monofunctional mechanism [6] or by a bifunctional one. Gates et al. [7] have reported that the bifunctional mechanism is controlled by the acid catalyzed step while Shum et al. [10] have found that the metal catalyzed step controls the reaction. Other workers [11, 12] also have noted that the dehydrocyclization on the metal function is important on the fresh catalyst but the acidic function becomes the controlling function after a few hours on stream due to coke deposition. On the other hand, the skeletal isomerization and cyclization of n-paraffins proceed through a multistep bifunctional mechanism as follows: n-alkanes are firstly dehydrogenated on metallic sites to the corresponding n-alkenes which are subsequently adsorbed on neighboring acidic sites and isomerized or cyclized to produce i-alkenes or cycloalkanes. The main reforming products (i.e., isoalkanes and aromatics) are finally produced on metallic sites by hydrogenation/dehydrogenation reactions of olefin and cyclic intermediates. It has been found that the rate determining step is the acid catalyzed one and thus the acidity is the rate controlling factor. Nevertheless, the metallic function is very important to determine the catalyst selectivity, and the regioselectivity of the first metallic step roughly determines the final product distribution. It has been proposed [13] that the equilibrium of the dehydrocyclization reaction is disfavored by a pressure increase since there is an increase of molecules number in the forward reaction. On the other hand, taking into account that the reaction is endothermic, an increase in the temperature favors the aromatics production. Furthermore, the thermodynamic feasibility of the reaction increases with the paraffin chain.

When the goal of naphtha reforming is to obtain aromatics, the paraffins dehydrocyclization is the most important reaction. The petroleum naphtha has a boiling point between 80 and 180°C and is composed of hydrocarbons chain in the range of C_5 to C_{12}. For the purpose of obtaining aromatics (benzene, toluene and xylenes), the naphtha cut should be in the range of C_6-C_8 [8]. For the production of xylenes, it is more interesting to reform n-octane by its direct dehydrocyclization or by dehydrocyclization of its isomers formed during the process. As shown in Figure 2, a mixture of ethylbenzene, *m*-xylene, *o*-xylene and *p*-xylene can be obtained in this case.

A simple reaction scheme proposed for n-octane reforming is shown in Figure 3. The n-paraffin is readily isomerized (IS) and also slowly dehydrocyclizated (DHC) to produce C8 aromatics. These compounds can also be obtained by the dehydrocyclization of C8 isomers. Some slow hydrocracking (HC) also occurs to produce lower molecular weight paraffins. The final products are aromatics and lower paraffins, some of them in the liquefied petroleum gas (LPG) range and methane.

Catalytic Naphtha Reforming Process

The catalytic naphtha reforming, an important refining and petrochemical process, was developed during the World War II to supply the demand of high octane gasoline. Currently, it is used not only to improve gasoline but also to produce high value aromatics, such as benzene, toluene and xylenes for the petrochemical industry. The production of aromatics or high octane gasoline depends on the feed and on the operation conditions. Hydrogen, a high value fuel and feedstock for petrochemical industries, is also obtained as byproduct.

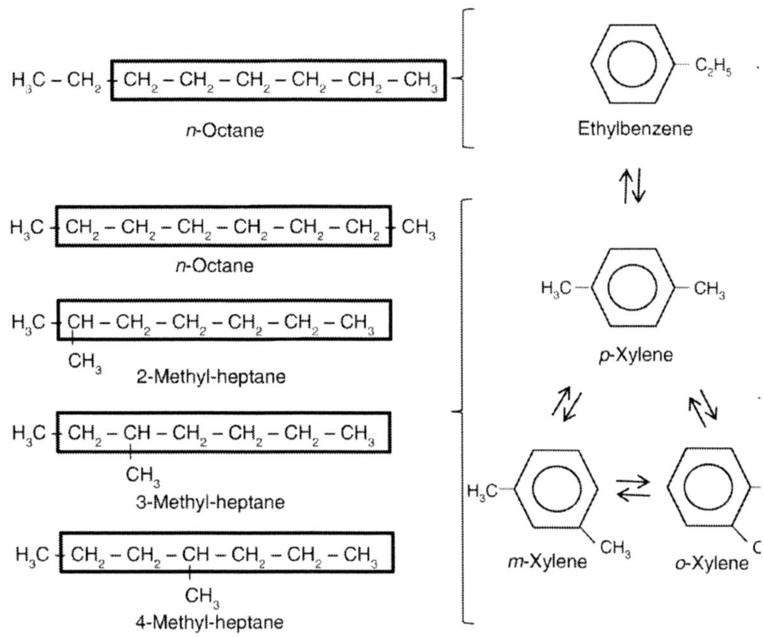

Figure 2. Scheme showing the closure of n-octane chain and its isomers to produce ethylbenzene, *m*-xylene, *o*-xylene and *p*-xylene by dehydrocyclization.

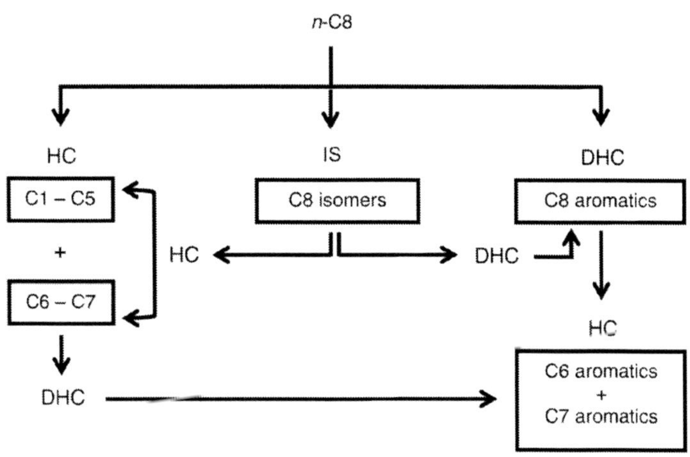

Figure 3. Scheme showing the reaction pathways in n-octane reforming.

An industrial unit of naphtha reforming is often made of three or four adiabatic fixed bed reactors, in series with heaters between them, to heat the effluent of the reactor before being fed to the following one. In the first reactor, the main reaction is the dehydrogenation of cycloalkanes to aromatics which is fast and endothermic, causing a decrease in temperature of 70 to 100°C. The residence time of the reactant as well as the reactor size and the catalyst mass are smaller as compared to the other ones. In the second reactor, the reaction rate is lower than in the first one, and the main reaction is the dehydroisomerization of naphthenes, which leads to a decrease in temperature of 22 to 32°C. The slowest reactions of the process, the dehydrocyclization and cracking occur in the third reactor. Due to the exothermicity of the hydrocracking reaction, there is a small temperature increase, which is beneficial to favor dehydrocyclization, thermodynamically favored by high temperatures. The liquid product of the process (reformate) comprises hydrocarbons in the range of five to ten carbon atoms, with 60-70 wt% of aromatics. The gas product contains hydrocarbons from one to four carbon atoms and 60-90 mol% of hydrogen. Part of the hydrogen produced is recycled to the process while the other one is consumed in hydrotreating processes in other units of the refinery [13-15].

The way and the frequency with which the regeneration of the naphtha reforming catalyst is performed determine the kind of the commercial catalytic reforming plants, which usually belong to one of the following types: semi-regenerative, cyclic or continuous regeneration (moving bed) [15]. A semi-regenerative unit is characterized by continuous operation for long cycles, with occasional stops for catalyst regeneration in order to restore its activity, which was lost by coke deposition. It is industrially conducted at high temperatures and moderate pressures (4-30 bar) [16]. The cyclic system differs from the semi-regenerative unit by an extra reactor (swing reactor), which allows the replacement of any of the other reactors for catalyst regeneration, while one remains in operation. This enables operation for longer cycles and in more severe conditions than in a semi-regenerative, i.e. lower pressures. However, the reactors work alternatively in reducing and oxidative atmospheres, requiring a complex design of the unit as well as rigorous safety precautions. In addition, the reactors must be of the same size, so that it can be possible the replacement between them [15, 16]. In the moving bed or continuous regeneration unit, the reactors are stacked on each other, and the catalyst moves slowly from the first reactor (top) to the bottom reactor. The coked catalyst removed from the last reactor is sent to the regeneration section and then returned to the top of the first reactor, keeping the total amount of catalyst [14-16]. Due to the operational conditions that prevent coke deposition on the catalyst, this arrangement has the advantage of being able to operate in more severe conditions (high temperatures and low pressures), keeping the content of coke in an acceptable level [15].

The main variables that affect the reforming process are temperature, pressure, hydrogen to hydrocarbon ratio and space velocity, which can affect the performance of the catalyst, changing the production and the quality of reformate [16, 17]. The operating conditions of the process reflect the thermodynamic balance between the undesirable and desirable reactions [18]. The desirable reactions are favored by high temperatures and low pressures, conditions that also favor coke production and thus the fast catalyst deactivation [17, 19]. As hydrogen reacts with the coke precursors removing them before they form significant quantities of polycyclic aromatics and deactivate the catalyst, the effect of the hydrogen to hydrocarbon ratio is equal to the pressure, i.e. by decreasing it there is an increase in the rate of coke formation with subsequent decrease of aromatics production [7, 17]. Furthermore, the space velocity has also to be controlled since low values can favor hydrocracking and other secondary reactions while high values can limit the life of the catalyst [17].

In spite of the control of these variables, during industrial operations the reforming catalyst undergoes several changes, which lead to progressive decline in catalyst performance decreasing the aromatics production. The temperature increase, a procedure often used to keep the activity in acceptable values during an industrial process, also leads to an increase in the rate of undesirable reactions. In addition, it causes a decrease in the desirable products yield and in the production of high pure hydrogen [12]. Furthermore, there is a limit to which the temperature may be elevated without affecting the structure of the catalyst. Another common procedure for keeping the catalyst activity is to add chloride compounds (hydrochloric acid, carbon tetrachloride or others) [19]. However, the addition of chloride also causes an increase in gas production, since it increases the support acidity favoring hydrocracking reactions. When the activity of the catalyst cannot be recovered anymore by increasing temperature and/or addition of chloride compounds, the end of the cycle is achieved and the catalyst must be regenerated [12].

The reforming catalysts deactivate mainly by coke formation [20]. The coke deposited on the active sites of both metal and support leads to a decrease in activity and to changes in the selectivity. The catalysts may also be deactivated by poisoning by sulfur, nitrogen, arsenic, sodium, copper, mercury, lead and silicium compounds [21-23]. The sulfur is the main poison which may deactivate both acidic and metallic sites [24].

Besides these conventional reforming technologies, some novel process have been introduced for aromatic production. These are the aroforming, M2-forming and cycler process, based on ZSM-5 and ZSM-5 containing gallium oxide catalysts [25].

First Generations of Naphtha Reforming Catalysts

The xylene contents in the reformate stream largely depend on the composition of feedstock, the type of reformer, the operating conditions and on the kind of the catalysts. These solids are especially important since by changing their properties, the yield of xylenes or even one of their isomers can be increased under the same process conditions [14, 16].

A great improvement in the naphtha reforming process occurred in 1949 when a bifunctional metal-acid catalyst was introduced replacing the former chromia-based ones. This solid gave rise to the so called platforming catalysts, which later has become the classical reforming catalyst. It was made of chlorided alumina-supported platinum and had both metallic and acidic sites which acted cooperatively for most reactions of naphtha reforming. The chlorided alumina was the acidic function which had enough acidity to catalyze the isomerization and cyclization steps while platinum was the best option to the metallic function, where hydrogenation and dehydrogenation occurred. The combined action of two sites promoted other reactions such as dehydrocyclization through a bifunctional mechanism [26, 27]. A typical industrial catalyst was made of 0.3 wt% platinum, with metallic area of around 0.5 $m^2.g^{-1}$ dispersed in acidified gamma alumina with high specific surface area (about 200 $m^2.g^{-1}$) [28].

In 1968, a new generation platforming catalysts based on alumina-supported platinum and rhenium was patented, showing that the addition of a second metal to the classical catalyst largely improved the performance of platinum. The bimetallic catalysts were more selective and stable than the platinum-based ones and enabled the process to run at a lower hydrogen pressure [29, 30].

The addition of rhenium to supported-alumina platinum catalyst led to a great improvement in its activity and stability, due to the decreasing amount of coke deposited as a result of a better removal of coke precursors [31]. However, since the first studies [16, 32] it was noted that the beneficial properties of rhenium addition required the sulfidation of the catalyst. This procedure involves the passivating of the catalyst with sulfur compounds (hydrogen sulfide or dimethyl sulphide) [16] to suppress the hydrogenolytic activity of rhenium which produces large amounts of gases. In this case, the sulfur atoms bind more strongly to rhenium and then protect platinum against poisoning. In addition, sulfur destroys the ensemble of platinum-rhenium atoms, which are responsible for hydrogenolysis [32].

After the introduction of rhenium to the platinum-based catalyst and its use in industrial processes, several other bimetallic systems were proposed and studied, such as those containing platinum and iridium [33], tin [34], tungsten or germanium, among others [35-43]. These new catalysts enable operations under different conditions making possible to obtain tailored reformates.

The main advantage of the bimetallic formulations, as compared to the monometallic ones, is a higher resistance against coke deactivation, allowing operations at more severe conditions (high temperatures and low pressures) and with longer cycles. Furthermore, these catalysts showed higher selectivities to the desired products than the monometallic one [28, 44]. Some of these catalysts also require pre-sulfidation such as platinum-iridium, which has high hydrogenolysis activity. Moreover, bimetallic catalysts based on tin and germanium do not need pre-sulphide since these metals are inactive to hydrogenolysis. Because of these advantages, these catalysts are promising candidates for naphtha reforming performed at low pressures, with continuous catalyst regeneration since it does not require complex procedures for activation [42].

In general, the role of the second metals is to modify some characteristics and properties of the metallic function such as dispersion, chemisorption capacity and activity for hydrogenolysis and dehydrogenation, among others. The acidic function also can be modified, such as strength and distribution of the sites, making them more suitable for a particular application as well as more selective to the desirable products. However, it is not completely clear how the second metal causes these changes in the platinum-based catalysts. It has been proposed [32] that the oxidation state of the metals may influence the catalytic properties, but this depends on numerous factors such as starting materials, methods for the catalyst preparation, metal contents, calcination and reduction procedures and kind of the support. These factors may determine, for instance, whether the promoters are present in the catalyst in reduced or oxidized form, forming alloys with platinum or interacting with the support.

The superiority of the bimetallic catalysts is often explained on basis of electronic and/or geometric effects of the metal promoter on platinum. These changes occur when platinum interacts with the oxides of the second metal or makes alloys with it. As a consequence, there is a weakening of the bond between platinum and the adsorbed hydrocarbon and the activity and selectivity change. The decrease in strength of the in Pt-C bond caused by the change in electron density of platinum atoms can explain the increased resistance of the metal against deactivation by coke deposition [28, 45].

Regarding the geometric effects, they become important when different structures are required on the catalyst surface. Some reactions such as hydrocracking, hydrogenolysis and coke formation, for instance, can occur only on particular arrangements of metal atoms (ensembles) and then are called *sensitive to structure* [46]. Moreover, other reactions, such as dehydrogenation, isomerization and aromatization, can occur over isolated platinum atoms (*insensitive to structure*) [45]. Therefore, the dilution of platinum atoms on catalyst surface by addition of tin or germanium, which are inactive in reaction, poisons selectively the active sites for hydrogenolysis, improving the formation of desired products and the resistance of the catalyst against coke [32]. In the case of systems based on platinum-rhenium and platinum-iridium, containing low amounts of sulfur, particles of rhenium-sulfur and iridium-sulfur are produced [28], which modify the surface structure of the catalyst similarly to tin or germanium.

The promoters can also affect the acidity of alumina usually used as support. Several authors [38, 47] have pointed out that tin is not completely reduced in the bimetallic catalyst and some tin oxide is formed and remains stabilized by the support. These species cause a decrease in the amount of acidic sites of the support, decreasing hydrocracking and increasing isomerization [38, 48-50]. Germanium oxide (GeO_2) in interaction with alumina also decreases the amount of acidic sites [51, 52].

As most heterogeneous catalysts, the properties of bimetallic catalysts largely depend on the preparation method, such as successive impregnation or coimpregnation, for instance. In addition, they depend on the kind of the second and/or third metal (rhenium, germanium, tin and others), the replacement of water by acetone or ethanol in the impregnation step and on the order of precursor addition, during impregnation. This makes possible to design a catalyst able to increase the xylenes yield as well as the increase of the production of some of their isomers.

Verderone et al. [53], for instance, have shown that an adequate balance between both metallic and acid functions is important to achieve high selectivities to aromatics hydrocarbons with suitable liquid yield and low coke deposition. This can be achieved by modifying the metal function by rhenium, tin, germanium or other metals while the acidity of the support (γ-Al_2O_3) can be tailored by the chlorine content, as well as by the oxide produced by the second metal (like SnO_2).

The physicochemical properties of the metallic phase, i.e., metal accessibility, interaction between the different metals and reduction degree are also influenced by the preparation method (coimpregnation, successive impregnation, catalytic reduction and others), the solvent used in impregnation (water or other solvent), the activation procedures (calcination, reduction), the chemical nature of metal precursors and the amount of metal. Both preparation and activation methods have strong influence on the physicochemical and catalytic properties of the resulting metallic phase. Therefore, the performance of naphtha reforming catalysts (i.e. activity, selectivity and stability) may be adequately modified using suitable preparation and activation techniques.

The kind of solvent used during the impregnation step is a typical preparation variable that greatly influences the performance of reforming catalysts. In a previous work [3], the effect of different solvents (water, acetone and ethanol) used in the impregnation step on the properties of Pt and Pt-Sn/Al_2O_3 catalysts for n-octane reforming was studied. It was found that *n*-octane produces mainly C_8 aromatics, such as *ortho*-, *meta*- and *para*-xylenes and ethylbenzene, besides benzene and toluene, at 500°C and atmospheric pressure, after 6 h on stream. The solvents led to catalysts with different properties, such as metal dispersion, amount of chloride, reduction behavior and metal particles size as well as different performances. These findings were related to the different boiling points and polarity of the solvents. Furthermore, the drying step was affected by the dipolar moment of the solvent, which has also a significant influence on the performance of the catalysts, mainly for n-pentane isomerization, cyclohexane dehydrogenation and n-octane reforming.

Figure 4 shows the selectivity of Pt and Pt-Sn/Al_2O_3 catalysts to xylenes, the most important products in reforming, at 1 and 6 h on stream. As aromatics are produced by paraffins dehydrocyclization through a bifunctional mechanism [13], the highest selectivity to xylenes was expected for the catalysts prepared using water or acetone, which exhibited the highest acidities. However, the highest selectivity to xylenes was found for the monometallic catalysts prepared using ethanol or acetone. This apparent discrepancy may be explained by considering the selectivities to hydrocracking (C_2-C_6) which are 9.7, 7.7 and 6.5 for the catalysts prepared using water, acetone or ethanol, respectively. Furthermore, the n-octane conversion was high (80-90 %) favoring the hydrocracking reactions, as found previously [13]. Therefore, the acidic function of these catalysts is able to catalyze hydrocracking reactions, decreasing the aromatics production. As the catalyst prepared in ethanol has the lower selectivity to hydrocracking, it produced more aromatics. This means that its acidity is not sufficiently high to decrease the aromatic production.

The higher xylenes selectivities for the bimetallic catalysts as compared to the monometallic ones can be related to their higher isomerization activity. The catalyst prepared in ethanol showed the highest acidity and xylene selectivity among the bimetallic catalysts. However, all bimetallic catalysts showed the same selectivity to hydrocracking (1.0%), which is lower than those showed by the monometallic catalysts, in agreement to the stronger acidities of the first ones. It means that tin affects the alumina strongest acid sites which are the most actives for hydrocracking reactions.

Last Generations of Naphtha Reforming Catalysts

The search for catalysts increasingly stable and selective to aromatics led to the addition of more than one metal to platinum, generating the multimetallic catalysts. Currently, the commercial operations use trimetallic catalysts with Pt-Re-M/Al_2O_3 type formulation.

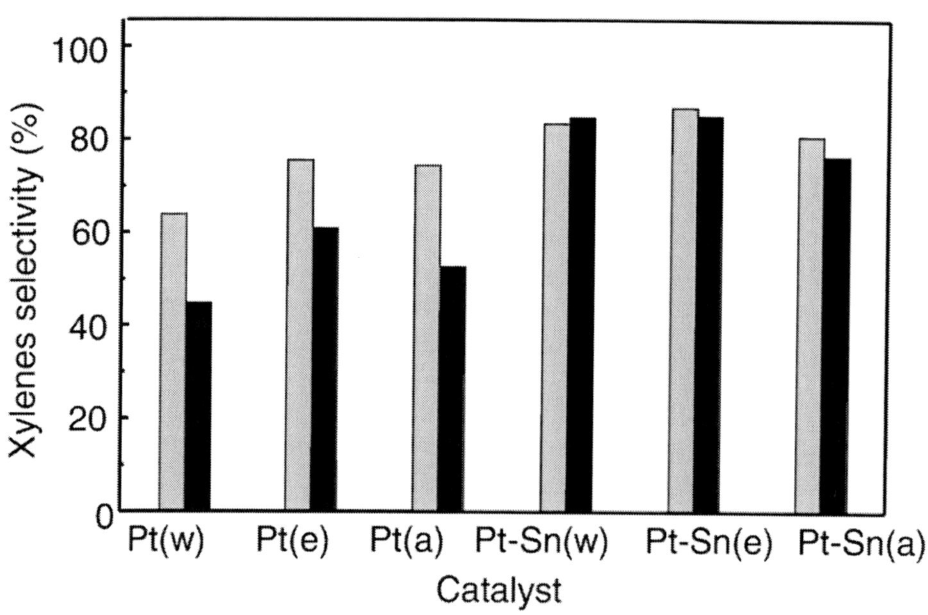

Figure 4. Xylenes selectivity at 1 (**grey**) and 6 h (**black**) on stream over Pt and Pt-Sn/Al_2O_3 catalysts during *n*-octane reforming. T = 500 °C; P = 1 MPa; WHSV = 1.5 h^{-1}; H_2/nC_8 (molar) = 19. w: water; e: ethanol; a: acetone.

In recent decades, several companies have implemented new alumina-supported catalysts in which platinum is promoted by two or more elements. Many patents involve trimetallic catalysts, such as Pt-Sn-In [54-57], Pt-Ir-Sn [58-61], Pt-Sn-Eu [62], Pt-Re-Ge [63, 64], Pt-Rh-Sn [65], Pt-Rh-In [66], Pt-Re-Ir [67], Pt-Co-Ga [68, 69], Pt-Re-In [70], Pt-Rh-Sn [71], which produce more aromatics and liquid products. Furthermore, other formulations of trimetallic reforming catalysts have been reported in open literature, such as Pt-Ir-Sn [72], Pt-Sn-M (M: Bi, Te, Au, Ir, Pd) [73, 74], Pt-Ir-Ge [75], Pt-Re-Sn [76, 77] and Pt-Re-Ge [78], among others.

For trimetallic catalysts, the metallic phase greatly affects the aromatics production. In a previous work, for instance, Fürcht et al. [74], have studied the n-octane reforming over alumina-supported Pt-Sn-Au, Pt-Sn-Ir and Pt-Sn-Pd catalysts at 10 atm and in the range of 480-510°C. They have found that the highest aromatic content were produced over Pt-Sn-Au and Pt-Sn-Ir catalysts, a fact that was ascribed to the ability of gold and iridium in producing smaller platinum ensembles. It is well known [79] that small platinum clusters have lower selectivity to hydrogenolysis, allowing the formation of desirable products. Gold and iridium-modified catalysts were slightly more active than the bimetallic platinum and tin-based one for n-octane reforming. On the other hand, palladium-modified catalysts produced lower amounts of aromatics and higher amounts of isoparaffins than the bimetallic one because palladium promoted cracking reactions. This finding was related to the ability of palladium in slightly deactivating the metal sites, which work properly in the bimetallic catalyst.

Other trimetallic catalyst, based on platinum, iridium and germanium, has been studied by Macleod et al. [75], who compared this solid to bimetallic catalysts containing these metals in n-octane reforming, using a pressurized microreactor system. They have found that tin and germanium modify the properties of these catalysts mainly through a geometric effect. Nevertheless, the formation of bulk Pt-Sn and Pt-Ge alloys contributed to the overall rate of deactivation of these systems. On the other hand, for the Pt-Ir and Pt-Re catalysts the resistance to deactivation by coking was improved, because of the ability of metallic rhenium and iridium to provide sites for the hydrogenolysis of coke deposits. Furthermore, the main effect of germanium on the trimetallic catalyst was to decrease the hydrogenolytic activity presumably due to a geometric effect. For the Pt-Ir-Ge catalyst, the production of aromatics was higher than that of platinum and tin bimetallic catalyst.

The effect of germanium on the platinum-based catalyst has also been studied by Santos et. al. [80], who have compared alumina-supported Pt-Re-Ge catalysts to the correspondent monometallic and bimetallic systems. It was noted that the metals formed alloys among them, and the electronic properties of platinum were affected by the other metals. The alloys were found to be inactive for dehydrogenation; this was assigned to the production of $Pt^{\delta+}$ or $Pt^{\delta-}$ species which were less active than Pt^0 species. However, the isomerization activity was increased while aromatization and hydrogenolysis ones were decreased. The production of o-xylene was increased by both rhenium and germanium. The trimetallic catalyst was the most active and stable one in *n*-octane reforming showing similar selectivities to aromatics to the germanium-based bimetallic catalyst.

Table 1 shows the selectivity of the catalysts to some products of *n*-octane reforming (naphthenes, aromatic, C_1, nC_2-nC_9, iC_4-iC_9). The selectivity to total aromatics was increased by rhenium addition while the isomerization was decreased. On the other hand, germanium addition reduced the total amount of aromatics and increased the isomerization products. The trimetallic catalyst showed similar selectivity to aromatics (46.7%) and to isomers (41.1%) as compared to the Pt-Ge bimetallic one but it was more stable. The addition of germanium and rhenium to platinum-based catalyst has caused opposite effects regarding the aromatics/isomers selectivities ratio, being the germanium more important than rhenium for both bimetallic and trimetallic catalyst. Table 1 also shows that Pt/Al_2O_3 and $Pt\text{-}Re/Al_2O_3$ catalysts have higher hydrogenolytic activities than the other catalysts. This is an important finding, since in commercial practice the undesirable hydrogenolytic activity is suppressed by catalyst sulfidation. The results showed indicate that germanium-containing catalysts do not require this procedure.

Table 2 shows the distribution of aromatics yield in reformate. One can see that germanium decreased ethylbenzene yield but greatly increased *o*-xylene yield.

Moreover, the addition of both rhenium and germanium (alone or together) increased mainly o-xylene formation, an important petrochemical feedstock.

The modification of $Pt\text{-}Sn/\gamma\text{-}Al_2O_3$ bifunctional catalyst by tin, bismuth and tellurium, prepared by metal adsorption technique, was also studied by Fürcht et al [73]. The catalysts were evaluated in n-octane reforming under similar conditions to industrial applications. The incorporation of tin slightly decreased the reformate yield while the aromatic content decreased and the isoparaffins content increased. However, the octane number remained high. Moreover, the addition of tellurium and bismuth seriously deactivated the metal function of the catalyst; the liquid yield of the product is lower than that over platinum–tin catalysts. However, the selectivity to isoparaffins and aromatic changed favorably. Regarding bismuth, it greatly deactivated the metallic function and considerably increased the cracking selectivity. Also, the reforming reactions which require the metallic function, without need of acidic centers (for example hydrogenolysis or dehydrocyclization), are hindered. Others reactions, which require alkenes and take place on the acidic function (with or without the contribution of the metallic sites), become more dominant among the possible reactions (isomerization, hydrocracking). The bismuth-doped catalyst showed almost zero cyclization selectivity, i.e. the product contained almost no aromatic and naphthenic compounds. It was argued that dehydrocyclization and aromatization are slow reactions as compared to isomerization, which occurs rapidly on acidic sites. The effect of tellurium on the activity and on the liquid yield was similar to bismuth, the metallic function was deactivated, and the liquid yield was lower than that of bimetallic platinum and tin catalyst but not so severe as happened for the bismuth-containing catalyst due to the fact that only a small fraction of tellurium remained close to platinum sites under operating conditions. The primary products of n-C_8 dehydrocyclization are ethylbenzene (EB) and *o*-xylene (OX). The formation of *m*-xylene (MX) and *p*-xylene (PX) needs the contribution of the acidic sites (isomerization of n-C_8 to methylheptanes and following dehydrocyclization to MX/PX or isomerization of EB/OX to MX/PX). Previously, Rangel et al. [40] and Parera and Fígoli [13] have reported that the OX/EB ratio is around 2 in thermodynamic equilibrium. Therefore, the high value (OX/EB= 4.3) for the tellurium-containing catalyst at 510°C can be explained by considering that there was a preferential 1,6-ring closure in *n*-octane through the participation of secondary carbon atoms rather than between primary and secondary carbons. Regarding the isomers, Fürcht et al [73] have proposed that the MPX/OXB improvement for the tellurium-based catalyst was due to an increase of isomerization selectivity. The formation and subsequent dehydrocyclization of 2-methylheptane produced the highest MPX/OXEB ratio.

Some preparation variables that are not expected to be important for the performance of reforming catalysts have shown, to significantly influence the catalytic behavior. One of the most important examples is shown by the effect of the order of addition of the metal precursors during impregnation [52, 77] on the properties of trimetallic catalysts. In studying the preparation of the Pt-Re-Sn/Al_2O_3 catalysts Carvalho et al. [77], for instance, successively impregnated the metals on the support with aqueous solutions in order to obtain a 0.3% of each metal in the final catalyst. It was found that the reduction of rhenium and tin to metals was affected by platinum. In addition, a large fraction of these metals remained oxidized and affected mainly the acidity of the support. When tin was added first, the acidic function was modified with positive consequences for the catalyst properties. Tin addition decreased the acidity that controls isomerization and cyclization reactions and decreased cracking and coking reactions because it blocks some strong acidic sites, keeping the moderate sites suitable for isomerization and cyclization.

The conversion of *n*-octane at 30 and 360 min on stream as well as the amount of carbon deposited on the catalysts, are shown in Figure 5 while the average values of selectivity to C_8 aromatics and to low value gases (C_1-C_4), are shown in Table 3. It can be noted an increase in activity, stability and selectivity to aromatics as well as a decrease in the selectivity to gases (C_1-C_4) due to rhenium or tin for bimetallic catalysts. Concerning the trimetallic catalysts, the best performance was shown by the solid prepared by adding tin before rhenium. The catalysts prepared by a reverse impregnation sequence (i.e. rhenium before tin) led to lower octane conversion, higher activity loss and coke deposition, lower selectivity to aromatics and higher production to gases. The best performance was shown by the Sn-Pt-Re catalyst, and the worst performance was presented by the Re-Pt-Sn one. This can be related to the fact that the dehydrocyclization reactions of *n*-C_8 are bifunctional, and controlled by the acidic function [7] whose amount and strength depend on the presence of metal oxide promoters. Therefore, the beneficial effect of tin can be related to the acidity of the support, as noted through n-pentane isomerization, this is closely related to the catalyst preparation.

The Sn-Pt-Re catalyst preparation begins with the impregnation of a solution of tin chloride on chlorided alumina, and tin ions are electrostatically adsorbed on specific sites of the support. After that, the catalyst is calcined and reduced. During the calcination step, the tin precursor is transformed into stannic oxide (SnO_2), which is heterogeneously bonded to alumina surface. From temperature programmed reduction, it was found that for the freshly calcined catalyst stannic oxide (SnO_2) was reduced to stannous oxide (SnO) in a broad temperature range (170-550°C) [76]. In addition, stannous oxide covers part of the alumina surface and strongly interacts with it forming tin aluminate ($SnAl_2O_4$). After platinum impregnation, the solid is dried, calcined and reduced again. The final reduction step transforms PtO_2 into Pt^0 species. The metallic platinum crystallites also catalyze the reduction of neighboring stannous oxide to Sn^0 which can be then alloyed to platinum. Therefore, the metal species which are expected to be found on the catalyst surface, are: both metallic Pt-Sn alloys and metallic platinum supported on alumina and metallic platinum supported on stannous oxide. The final deposition of rhenium on the support has given more stability to the final catalyst as compared to the bimetallic Pt-Sn one. Depending on the degree of hydration of the support during reduction, the surface rhenium oxide clusters become mobile and can migrate to platinum crystals where they are reduced. As a consequence, both Pt-Re and Pt-Sn-Re aggregates were produced. Rhenium oxide clusters interact with alumina as well as with stannous oxide and then are stabilized on them. For this reason, the catalyst surface is supposed to be formed by crystallites of Pt, Pt-Sn, Pt-Re, Re-Sn and Pt-Sn-Re supported on alumina.

Table 1. Catalysts selectivity to some products during the *n*-octane reforming at 5 and 360 min on stream. T = 450°C; P = 1MPa; WHSV = 3.4; H_2/nC_8 (molar) = 7.3

Catalyst	Pt		Pt-Re		Pt-Ge		Pt-Re-Ge	
Time on stream (min)	5	360	5	360	5	360	5	360
C_1	0.8	0.4	0.4	0.3	0.2	0.1	0.2	0.2
nC_2-nC_9	13.7	17.3	12.1	13.2	10.8	11.0	10.7	9.7
iC_4-iC_9	34.6	61	31.4	48.2	42.2	51.9	41.1	51.8
Naphthenes	2.6	1.3	2.0	1.5	1.1	2.1	1.0	2.0
Total aromatics	47.1	19.4	53.4	36.3	45.4	34.6	46.7	36.1
Coke (%)	-	0.53	-	0.49	-	0.42	-	0.41

Table 2. Aromatic yield for monometallic, bimetallic and trimetallic catalyst during the *n*-octane reaction at 5 min on stream

Catalysts	Pt	PtRe	PtGe	PtReGe
Benzene	0.4	0.6	0.2	0.2
Toluene	2.7	2.5	0.8	0.9
EB	6.6	7.1	3.7	4.0
PX	2.6	3.1	2.6	2.7
MX	7.5	8.5	7.3	8.0
OX	6.4	9.8	10.0	10.6

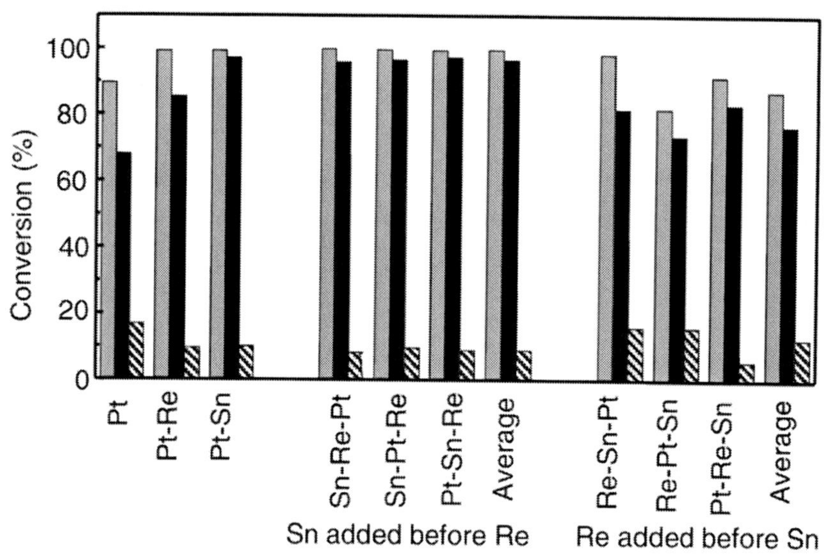

Figure 5. *n*-Octane conversion at 30 (▭) and 360 (▬) min on stream and amount of coke (▨) deposited on the catalysts. Reaction conditions: 500°C; 1 atm, WHSV = 1.8; H_2/n-C_8H_{18} (molar) = 10.

The trimetallic catalyst which has shown the worst performance was obtained when rhenium was added before platinum and then tin in this order. This can be related to rhenium oxidized species which strongly interact with the support during the first calcination step. When platinum and tin precursors were added after rhenium, the most active adsorption sites on alumina could have been at least partially blocked by the previously formed oxidized rhenium species and then platinum and tin could interact more freely. After calcination, the finished catalyst should have tin and platinum in strong interaction so the effect of tin on support acidity should be negligible. The amount and strength of acidic sites for this trimetallic catalyst should be similar to that of the Pt-Re bimetallic one and the metallic phase should not be very different from Pt-Sn phase for bimetallic catalyst. The addition of tin prior to Re affected the acidic function of the catalyst favoring isomerization and cyclization and decreasing cracking. It is well known [7] that paraffin isomerization is a very fast reaction while cracking and cyclization are slower ones.

Considering the activity, selectivity and stability, the most suitable catalyst for n-octane dehydrocyclization is the Sn-Pt-Re one. When tin precursor was firstly added, it was strongly adsorbed on alumina, and the beneficial effects on acidity were observed. Tin decreased the support acidity by blocking some strong acidic sites responsible for cracking and coking while moderate sites suitable for isomerization and cyclization remain. When rhenium was added later, the most affected catalytic function was the metallic one by decreasing the platinum crystallites size. The incorporation of rhenium at first increased the support acidity and then the selectivity and stability were decreased. It was concluded that the main advantage of trimetallic catalysts is the increase in selectivity and stability, allowing the possibility of tailoring catalysts for increasing the xylenes yield.

The influence of the order addition of precursors on Pt-Re-Ge/Al_2O_3 catalysts was also studied [52]. It was found that the catalysts properties in n-octane reforming are strongly affected by this variable. The impregnation of platinum prior to rhenium and germanium decreased the interaction between the metals, leading to catalysts considerably more active in cyclohexane dehydrogenation and less active in hydrogenolysis, as compared to bimetallic and monometallic catalysts. The incorporation of germanium before the other metals led to catalysts with lower hydrogenation activity and higher hydrogenolytic activity, due to strong metal interactions. The catalysts obtained by incorporating rhenium before the other components have shown an intermediate behavior. The metallic phase properties were more affected by the presence of rhenium than by germanium. The acidic properties of the support were not modified by the order addition of metal precursors. However, germanium and rhenium changed the support acidity in the surrounding of the deposited metal and then modified the selectivities to aromatics and paraffinic isomers. It also noted that the selectivity to isomers decreased as the selectivity to aromatics increased. Furthermore, the selectivity to isoparaffins or aromatics was drastically changed while the activity slightly varied with the order addition of the precursors. On the other hand, the global selectivity to isoparaffins and aromatics remained nearly constant regardless the order of addition of the precursors. According to the current accepted reaction mechanism for catalytic reforming, olefins are first produced on metal sites and then they are isomerized or transformed over the acid sites [13], as shown in Figure 3. These reactions on the acidic sites are the slowest ones, so they are rate-controlling. As the rate of aromatics production was higher than the rate of isoparaffins production it can be concluded that the interaction between platinum and rhenium is higher in the Pt-Re-Ge and Pt-Ge-Re catalysts. This is an evidence that germanium and rhenium oxides in the surrounding of platinum modified the acidity of the support and favor dehydrocyclization. These characteristics can be used to tailor catalysts with high selectivity for aromatics or isoparaffins, depending on the purpose of the reforming process. The Ge-Re-Pt catalyst showed the highest selectivity to aromatics.

Table 3. Selectivity of alumina-supported platinum and/or rhenium and/or tin catalysts prepared by successive impregnations in *n*-octane on presulfided catalysts. Reaction conditions: 500°C; 1 atm; WHSV = 1.8; $H_2/n\text{-}C_8H_{18}$ (molar)= 10

Catalyst	Sel. arom C_8 (%)	Sel. C_1-C_4 (%)
Pt	43	23
Pt-Re	73	11
Pt-Sn	89	5
	Sn added before Re	
Sn-Re-Pt	86	3.6
Sn-Pt-Re	87	3.2
Pt-Sn-Re	87	3.9
Average	87	3.6
	Re added before Sn	
Re-Sn-Pt	80	6.2
Re-Pt-Sn	74	6.5
Pt-Re-Sn	72	6.1
Average	75	6.3

Conclusion

Catalytic naphtha reforming is the main commercial route to produce high value aromatics, especially xylenes, which have found several applications in different fields. This process was developed during the World War II and went on several improvements over the years. The main one was the addition of metals to the classical alumina-supported platinum, giving rise to the multimetallic catalysts. Nowadays, the catalytic naphtha reforming is a mature technology in which trimetallic catalysts are often used. However, in spite of the high performance of these catalysts, several studies have still carried out aiming to improve those solids even more. The most used formulation is based in alumina-supported platinum and rhenium containing tin or germanium.

References

[1] Hocking, MB. *Handbook of Chemical Technology and Pollution Control*, third ed., Academic Press, New York, 2005.

[2] Dias, JA; Rangel, MC; Caliman, E; Garcia, FAC ; Dias, SL. *Appl. Catal. A: Gen.* 328, (2007), 189-194.

[3] Alabi, W; Atanda, L; Jermy, R; Al-Khattaf, S. *Chem. Eng. J.* 195–196, (2012), 276-288.

[4] Tsai, TC; Liu, SB; Wang, I. *Appl. Catal. A: Gen.* 181, (1999), 355-398.

[5] Breen, JP; Burch, R; Kulkarni, M; McLaughlin, D; Collier, PJ; Golunski, SE. *Appl. Catal. A:Gen.* 181, 316, (2007), 53-60.

[6] Garin, F; Gault. FF. in: R. Prins and G. C. A. Schuit, (Eds), Chemistry and Chemical Engineering of Catalytic Processes, Sijthoff and Noordhoff, *Alphen aan den Rijn*, 1980, 351-380.

[7] Gates, BC; Katzer, JR; Schuit, GCA. in: R. Ciofalo, D. J. Marshall, B. Leap, (Eds.), *Chemistry of Catalytic Processes*, New York: McGraw-Hill, 1979, 184-324.

[8] Prestvik, R; Moljord, K; Grande, K; Holmen, A. in: G. J. Antos and A. M. Aitani, (Eds.), *Catalytic Naphtha Reforming*, second ed., Marcel Dekker Inc., New York, 2004, 1-33.

[9] Murthy, KR; Sharma, N; George, N. in: G. J. Antos, A. M. Aitani, J. M. Parera, (Eds.), Catalytic Naphtha Reforming. *Science and Technology*, Marcel Dekker, Inc., New York, 1995, 207-255.

[10] Shum, VK; Butt, JB; Sachtler, WMH. *Appl. Catal. A: Gen.* 11, (1984), 151-154.

[11] Querini, CA; Figoli, NS; Parera, JM. *Appl. Catal. A: Gen.* 52, (1989), 249-262.

[12] Parera, JM; Fígoli, NS. *Catalysis, Royal Society Chemistry*, Cambridge, 1992.

[13] Parera, JM; Fígoli, NS. in: G. J. Antos, A. M. Aitani, J. M. Parera, (Eds.), Catalytic Naphtha Reforming. *Science and Technology*, Marcel Dekker, New York, 1995. 45-77.

[14] Little, DM. *Catalytic Reforming*, Rennwell Publishing Company, Oklahoma, 1985.

[15] Aitani, AM. in: G. J. Antos, A. M. Aitani, J. M. Parera, (Eds.),. Catalytic Naphtha Reforming. *Science and Technology*, Marcel Dekker Inc., New York, 1995. 409-436.

[16] Edgar, MD. in: B. E. Leach, (Ed.). *Applied Industrial Catalysis*, Academic Press, New York, 1983, 123-148.

[17] Figoli, NS; Beltramini, JN; Martinelli, MRS ; Parera, JM. *Appl. Catal. A: Gen.* 5, (1983), 19-32.

[18] Ciapetta, DN; Wallace, DN. *Catal. Rev.-Sci. Eng.* 5, (1971), 67-158.

[19] Parera, JM; Beltramini, JN. *J. Catal.* 112, (1988), 357- 65.

[20] Breen, J; Burch, R; Collier, P; Golunsk, S. *J. Am. Chem. Soc.* 127, (2005), 5020-5021.

[21] Beltramini, JN. in: G. L. Antos, A. M. Aitani, J. M. Parera, (Eds.), Catalytic Naphtha Reforming. *Science and Technology*, Marcel Dekker, New York, 1995. 313-363.

[22] Souza, MOG; Reyes, P; Rangel, M. *Stud. Surf. Sci. Catal.* 126, (1999), 469-472.

[23] Barbier, J; Marecot, P; Tifouti, L; Guenim, M; Fretty, R. *Appl. Catal. A: Gen.* 19, (1985), 375-383.

[24] Oudar, J. *Catal. Rev. – Sci. Eng.* 22, (1980), 171-95.

[25] Sivasanker, S; Ratnasamy, P. in: G. J. Antos, A. M. Aitani, J. M. Parera, (Eds.),, Catalytic Naphtha Reforming. *Science and Technology*, New York, Marcel Dekker, 1995. 483-507.

[26] Haensel, V. UOP, *U. S. Patent*, 2, 479, 109, (1949).

[27] Haensel, V. UOP, *U. S. Patent*, 2, 479, 110, (1949).

[28] Ribeiro, FH; Bonivardi, AL; Kim, C; Somorjai, GA. *J. Catal.* 150, (1994), 186-198.

[29] Klusksdahl, HE. *U. S. Patent*, 3, 415, 737, (1968).

[30] Saterfield, CN. *Heterogeneous Catalysis in Practice*, McGraw–Hill, Boston, 1980.

[31] Parera, JM; Verderone, RJ; Querini, CA. *Stud. Surf. Sci. Catal.* 34, (1987), 135-145.

[32] Macleod, N; Fryer, JR; Stirling, D; Webb, G. *Catal. Today*, 46, (1998), 37-54.

[33] Sinfelt, JH; *U. S. Patent*, 3, 953, 368, (1976).

[34] Weisang, JE; Engelhard, P. *French Patent*, 2, 031, 984, (1969).

[35] McCallister, KR; O'Neal, TP. *Catalyseur*, French Patent, 2, 078, 056, (1971).

[36] McVicker, GB; Ziemiak, JJ. *Appl. Catal. A: Gen.* 14, (1985), 229-243.

[37] Rice, RW; Lu, K. *J. Catal.* 77, (1982), 104-117.
[38] Carvalho, LS; Reyes, P; Pecchi, G; Fígoli, NS; Pieck, CL; Rangel, MC. *Ind. Eng. Chem. Res.* 40, (2001), 5557–5563.
[39] González-Marcos, MP; Iñarra, B; Guil, JM; Gutiérrez-Ortiz, MA. *Catal. Today*, 107–108, (2005), 685–692.
[40] Rangel, MC; Carvalho, LS; Reyes, P; Parera, JM; Fígoli, NS. *Catal. Lett.* 64, (2000), 171-178.
[41] Contreras, JL; Del Toro, G; Schifter, I; Fuentes, GA. in: J. W. Ward, (Ed.), *Catalysis*, Elsevier Science Publishers B. V., Amsterdam. 1988. 51-59.
[42] Borgna, A; Garetto, TF; Apesteguía, CR; Moraweck, B. *Appl. Catal. A: Gen.* 182, (1999), 189-197.
[43] Mariscal, R; Fierro, JLG ; Yori, JC; Parera, JM; Grau, JM. *Appl. Catal. A: Gen.* 327, (2007), 123-131.
[44] Verderone, KJ; Pieck, CL; Sad, MR; Parera, JM. *Appl. Catal. A: Gen.* 21, (1986), 239-250.
[45] Coq, B; Figueras, F. *J. Catal.* 85, (1984), 197-205.
[46] Boudart, M; Aldag, A; Benson, JE; Dougharty, NA; Harkins, CG. *J. Catal.* 6, (1966), 92-99.
[47] Burch, R; Garla, LC. *J. Catal.* 71, (1981), 360-372.
[48] Sparks, DE; Srinivasan, R; Davis, BH. *J. Mol. Catal.* 88, (1994), 325-341.
[49] Sparks, DE; Srinivasan, R; Davis, BH. *J. Mol. Catal.* 88, (1994), 359-376.
[50] Srinivasan, R; Davis, BH. *J. Mol. Catal.* 88, (1994), 343-358.
[51] Boutzeloit, M; Benitez, VM; Mazzieri, VA; Especel, C; Epron, F; Vera, CR; Pieck, CL; Marecót, P. *Catal. Comm.* 7, (2006), 627-632.
[52] Carvalho, LS; Conceição, KCS; Mazzieri, VA; Reyes, P; Pieck, CL; Rangel, MC. *Appl. Catal. A: Gen.* 419– 420, (2012), 156–163.
[53] Verderone; RJ; Pieck; CL; Sad, MR; Parera, JM. *Appl. Catal.* 21, (1986), 239-250.
[54] Bogdan, PL; Imai, T. *U. S. Patent*, 6, 048, 449, (2000).
[55] Bogdan, PL. *U. S. Patent*, 6, 013, 173, (2000).
[56] Robinson, DW; Johnson, RW; Moser, MD. *U. S. Patent*, 4, 487, 848, (1984).
[57] Bogdan, PL; Imai, T. *U. S. Patent*, 5, 858, 908, (1999).
[58] Baird, WC; Boyle, JP; Swan III, GA. *U. S. Patent*, 5, 269, 907, (1993).
[59] Baird, WC; Boyle, JP; Swan III, GA. *U. S. Patent*, 5, 106, 809, (1992).
[60] Bogdan, PL. *U. S. Patent*, 6, 495, 487(2002).
[61] Bogdan, PL. *U. S. Patent*, 6, 884, 340, (2005).
[62] Bogdan, PL. *U. S. Patent*, 5, 665, 223, (1997).
[63] Antos, GJ. *U. S. Patent*, 4, 312, 788, (1982).
[64] Kresge, CT; Krishnamurtly, S; McHole, WD. *U. S. Patent*, 4, 493, 764, (1985).
[65] Moser, MD; Lawson, RJ. *U. S. Patent*, 4, 737, 483, (1988).
[66] Antos, GJ. *U. S. Patent*, 4, 149, 961, (1976).
[67] Baird Jr., WC. *U. S. Patent*, 4, 746, 418, (1988).
[68] Sorrentino, CM; Pellet, RJ; Bertolacini, RJ. *U. S. Patent*, 4, 482, 449, (1984).
[69] Sorrentino, CM; Pellet, RJ; Bertolacini, RJ. *U. S. Patent*, 4, 469, 812, (1984).
[70] Antos, GJ; Wang, L. *U. S. Patent*, 4, 629, 551, (1986).
[71] Moser, MD; Lawson, RJ. *U. S. Patent*, 4, 714, 538, (1987).
[72] Carnevillier, C; Epron, F; Marecot, P. *Appl. Catal. A: Gen.* 275, (2004), 25-33.

[73] Fürcht, A; Tungler, A; Szabó, S; Sárkáni, A. *Appl. Catal. A: Gen.* 226, (2002), 155-161.
[74] Fürcht, A; Tungler, A; Szabó, S; Schay, Z; Vida, L; Gresits, I. *Appl. Catal. A: Gen.* 231, (2002), 151-157.
[75] Macleod, N; Fryere, JR; Stirling, D; Webb, G. *Catal. Today*, 46, (1998), 37.
[76] Carvalho, LS; Pieck, CL; Rangel, MC; Fígoli, NS; Grau, JM; Reyes, P; Parera, JM. *Appl. Catal. A: Gen.* 269, (2004), 91-103.
[77] Carvalho, LS; Pieck, CL; Rangel, MC; Fígoli, NS; Vera, CR; Parera, JM. *Appl. Catal. A: Gen.* 269, (2004), 105-116.
[78] D'Ippolito, SA; Vera, CR; Epron, F; Especel, C; Marécot, P; Pieck, CL. *Catal. Today*, 133–135, (2008), 13-19.
[79] Mazzieri, VA; Grau, JM; Yori, JC; Vera, CR; Pieck, CL. *Appl. Catal. A: Gen.* 354, (2009), 161-168.
[80] Santos, MCS; Grau, JM; Pieck, CL; Parera, JM; Fierro, JLG ; Fígoli, NS; Rangel, MC. *Catal. Lett.* 103, (2005), 229-237.

In: Xylenes
Editor: Michael Olawale Daramola

ISBN: 978-1-62808-342-2

Chapter 3

Xylenes Production from Renewable Feedstocks

***T. Sooknoi*[1] *and A. Ausavasukhi*[2]**
[1]Department of Chemistry, Faculty of Science, King Mongkut's Institute of Technology Ladkrabang, Bangkok, Thailand
[2]Programme in Applied Chemistry, Faculty of Sciences and Liberal Arts, Rajamangala University of Technology Isan, Nakhon Ratchasima, Thailand

Abstract

Considerable efforts have been made during the past few years for utilizing renewable resources for chemical feedstock. As being ligno-cellulosic materials, biomass provides ready-to-use aromatic precursors for the production of xylene from biomass-derived compounds. Typical approach involve fast-pyrolysis of the biomass followed by catalytic deoxygenation of the aromatic components (mainly the phenolics) in the "pyrolysis bio-oil" As the pyrolysis products always possess reactive oxygenates such as alcohols, aldehydes, and acids, the reaction requires a catalyst that readily promotes deoxygenation and/or dealkylation to form less substituted aromatic ring. Metals supported acidic oxides have been widely investigated for the reactions involving hydrogen (hydrogenation/ dehydrogenation and hydrogenolysis) and carbocation intermediates. Alternatively, refined biomass such as ethanol, fatty acid and glucose can also be converted to aromatic hydrocarbons. This approach involves dehydration, decarbonylation and/or decarboxylation to primarily produce olefinic or paraffinic hydrocarbons which can be subsequently converted to aromatic via classical acid catalysed processes. Shape selective feature of the zeolites always plays marked role for selective formation of xylenes for all approaches. This chapter will review recent attempts for xylene productions from renewable feedstock. The relation between the molecular architecture of the biomass-derived compounds and the selected catalyst system will be highlighted, together with the reaction conditions and possible on-going xylene production processes.

Keywords: Xylenes, aromatics, renewable, pyrolysis

INTRODUCTION

Mononuclear aromatic compounds (benzene, toluene, ethylbenzene, and xylenes abbreviated BTEX) are one of the most important chemical feedstocks widely used as the raw material of monomers, paints, detergents, dyes, pesticides, and pharmaceuticals. The BTEX present in crude oil are generally low in concentration and not technically or economically feasible to separate them. However, an aromatic-rich mixture can be obtained from the refinery streams, especially from catalytic reforming and cracking, and pyrolysis gasoline from steam cracking and from coal liquids.

Conversion of lighter and heavier aromatics to C_8 aromatics (xylenes and ethylbenzene) is becoming an essential part of modern integrated refinery-petrochemical complex due to increased demand of xylenes. A list of key reactions is presented in Table 1.

However, such efforts are developed for the production of xylene from petroleum-based resources. It is better if all the raw materials of chemical industries and portions of other needs can be shared by biomass-derived compounds. Because of factors such as the finite reserves of petroleum, their unequal geographic distributions, and the environmentally motivated desire to decrease CO_2 emissions, renewable feedstocks, such as ligno-cellulosic biomass, have become promising candidates for the production of renewable fuels and chemicals. Although catalytic conversion of biomass-derived compounds can directly produce xylenes, such process can also generate lighter and heavier aromatics. Therefore, upgrading of lighter and heavier aromatics to xylene preferably involves disproportionation, transalkylation, alkylation, and isomerization (as shown in Table 1). The production of xylenes from the biomass-derived compounds is probably the most economically favorable process. This chapter will review recent attempts for xylene productions from renewable feedstock including the conversion of (i) ethanol, (ii) ester and fatty acid, (iii) pyrolysed product, and (iv) glucose to BTEX aromatics.

Table 1. Xylenes production from lighter and heavier aromatics

Reaction	Reactants	Products	Catalyst	Conditions	Ref.
Disproportionation	Toluene + Toluene	Xylenes + Benzene	HZSM-5, MCM-22, Composite catalyst: Borosilicate/MCM-22	T = 200-360°C	[1-3]
	C_9 aromatics + C_9 aromatics	Xylenes + C_{10} aromatics	HBeta, HY, USY-based FCC	T = 350-500°C	[4-5]
Transalkylation	Toluene + C_9 aromatics	Xylene + Xylene	Silica-supported $H_3PW_{12}O_{40}$ (H_3PW), HBeta	T = 300-470°C	[6-7]
Alkylation	Toluene + Methanol	Xylenes	ZM13, HZSM-5, Mordenite/ZSM-5, TNU-9, MCM-22, Pt-modified nano-scale HZSM-5	T = 300-460°C	[8-10]
Isomerization	Xylene	*p*-Xylene	ZSM-5, mordenite, SSZ-33, TNU-9, Pt-HZSM-5, Composite catalyst: ZSM-5/MCM-48, ZSM-5/MCM-41	T = 300-400°C	[11-13]

CONVERSION OF THE PROLYSIS BIO-OIL

Fast pyrolysis of biomass followed by catalytic conversion of the pyrolysed products is of particular interest, as it is the process that produces xylene from ligno-cellulosic biomass. Literally, pyrolysis corresponds to the thermal conversion occurring in the absence of oxygen. Biomass pyrolysis results in the production of three products: gas, pyrolysis oil and charcoal. The relative proportions of these products depend very much on the pyrolysis methods, the characteristics of the biomass and the reaction parameters. Temperature, heating rate and vapour residence time can be adjusted to favor liquid productions. As the pyrolysis products always possess reactive oxygenates such as alcohols, aldehydes, and acids, the reaction requires a catalyst that readily promotes deoxygenation and/or dealkylation to form less substituted aromatic ring. The major aromatic components from the pyrolysis are composed mainly of cresol, guaiacol, anisole and benzaldehyde. According to their different functionality the deoxygenation process of these components can be ascribed in the following sections.

The Production of BTEX Aromatics from Aromatic Alcohols (m-Cresol and Guaiacol)

As C-O bonding of phenolic compound is relatively strong, a catalyst with oxophillicity center is required for hydrogenolysis of such particular bond. Efforts in the xylene production are focused towards selective hydrodeoxygenation over Ga loaded catalyst. Ga-modified HBeta (Ga/HBEA) zeolites were evaluated and compared to other Ga-containing catalysts (i.e. SiO_2 and ZSM-5 (MFI)) for the hydrodeoxygenation of *m*-cresol, a model compound representative of lignin-derived phenolics [14].

The conversions of *m*-cresol are conducted at 400 and 450°C, atmospheric pressure, 6.3 W/F (h), and using H_2 as carrier gas. Table 2 shows that at 400°C, oxygenated compounds are the dominant products (~50 wt.%) over 1Ga/HBeta. However, the increase in Ga loading enhanced formation of benzene, toluene, and xylene (~33 wt.%). It is expected that the C-C and C-O hydrogenolysis activity is provided by the incorporation of Ga in the HBeta zeolite.

Scheme 1 summarizes the possible products and reaction pathways for *m*-cresol conversion. A surface pool of oxygenates would be the primary precursors for phenol and the observed heavy oxygenated compounds. They can also be hydrodeoxygenated, producing the observed light aromatics (benzene, toluene, xylene) and light hydrocarbons (C_2-C_6 hydrocarbons).

The selectivity to benzene, toluene, and xylene increased with temperature in the range of 400-550°C, while that to phenol and oxygenated compounds decreased. In fact, this is an expected trend from C-O and C-C hydrogenolysis of phenol and other oxygenated compounds since these reactions have relatively high activation energies.

Table 2. The *m*-cresol conversion over Ga-modified HBeta (Ga/HBEA) zeolites [14]

Beta zeolite	H	H	1Ga/H	3Ga/H	6Ga/H
Temperature (°C)	450	400	400	400	400
% Conversion	70.45	55.51	70.76	70.56	71.64
% Yield					
Methane	0.70	0.87	1.43	1.32	1.55
Benzene	0.80	0.68	3.24	3.45	6.12
Toluene	1.59	2.90	8.74	9.45	21.67
Xylene	0.54	1.38	3.56	3.20	5.40
Other hydrocarbon	2.45	2.83	3.75	3.30	4.80
Phenol	18.22	32.00	19.76	23.61	24.66
Oxygenated compound	46.16	14.83	30.28	26.23	7.44
Total yield	70.45	55.51	70.76	70.56	71.64
% Selectivity					
Methane	0.99	1.58	2.02	1.87	2.17
Benzene	1.14	1.23	4.58	4.89	8.54
Toluene	2.25	5.23	12.35	13.39	30.25
Xylene	0.76	2.49	5.03	4.54	7.53
Other hydrocarbon	3.47	5.10	5.30	4.67	6.70
Phenol	25.86	57.65	27.93	33.45	34.42
Oxygenated compound	65.51	26.71	42.79	37.18	10.39
Total selectivity	100.00	100.00	100.00	100.00	100.00

Reaction condition : W/F = 6.3 h, Reaction temperature = 400 and 450°C, Carrier gas = H_2, Pressure = 1 atm.

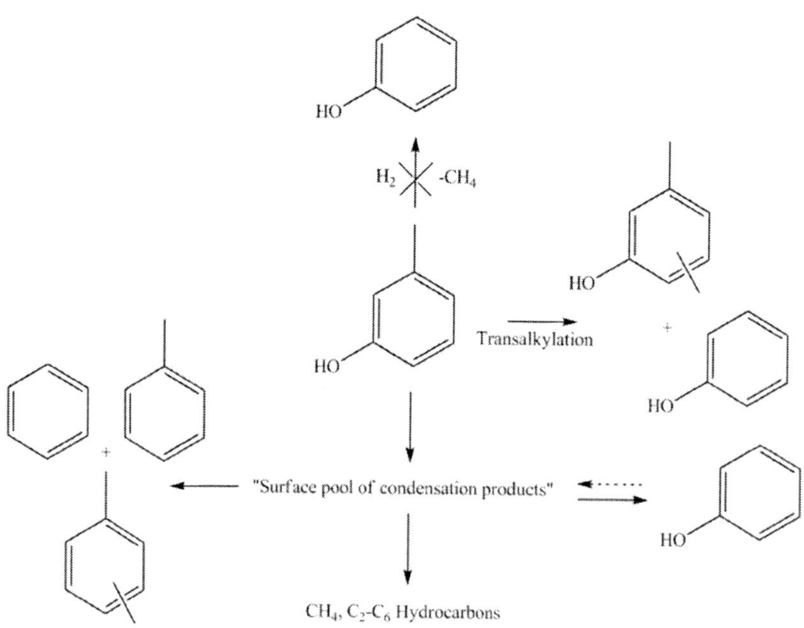

Scheme 1. Reaction pathway for the conversion of *m*-cresol to aromatics, phenol, and trapped surface species [14].

Table 3 compares the product yields obtained over the different catalysts under the same reaction conditions. Clearly, 3Ga/HBeta is more active than 3Ga/HZSM-5 and much more than 3Ga/Silica. The low conversion and low selectivity to light aromatics obtained on 3Ga/Silica can be explained by 2 reasons. First, silica is not an effective support for stabilizing highly dispersed Ga species (Ga^+/GaH_2^+). Second, silica does not contain any significant acidity, which seems to play a role in the conversion of cresol, as evidenced by the relatively high activity exhibited by the Ga-free zeolites (HBeta and HZSM-5). While, the observed difference in the activity of the two zeolites is most likely related to their different structure. Diffusion of a rather bulky molecule such as *m*-cresol may be inhibited within the restricted pores of ZSM-5, as compared to that in the larger pores of Beta zeolite. It is interesting to note that shape selective feature of the zeolites always plays marked role for selective formation of xylenes.

Guaiacol (2-methoxyphenol), the model molecules representative aromatic alcohol from ligno-cellulosic biomass liquefaction, was also studied over unsupported and alumina-supported MoS_2 and CoMoS catalysts [15].

The conversions of guaiacol were conducted at 300°C and the total hydrogen pressure was 4 MPa. The reactant was introduced with H_2 flow through a saturator/condenser system at a partial pressure of 2.7 kPa. The basic reaction scheme of guaiacol (GUA) hydrodeoxygenation involves transformation into catechol (CAT), phenol (PHE) and deoxygenated compounds (benzene, cyclohexene, cyclohexane) as well as methyl-substituted compounds (Scheme 2).

Table 3. Effect of support material [14]

Type of catalyst	3Ga/HBeta	3Ga/Silica	3Ga/HZSM-5	HZSM-5
% Conversion	82.78	4.49	40.32	41.69
% Yield				
Methane	2.25	0.78	0.34	1.50
Benzene	10.45	0.15	1.24	1.24
Toluene	23.06	0.32	8.90	2.00
Xylene	9.84	0.00	0.86	0.00
Other hydrocarbons (C_2-C_6)	1.83	0.00	0.90	1.50
Phenol	13.78	2.82	19.91	23.90
Oxygenated compounds	21.57	0.41	8.18	11.55
Total yield	82.78	4.49	40.32	41.69
% Selectivity				
Methane	2.72	17.45	0.84	3.60
Benzene	12.62	3.40	3.08	2.97
Toluene	27.85	7.19	22.06	4.80
Xylene	11.88	0.00	2.13	0.00
Other hydrocarbons (C_2-C_6)	2.22	0.00	2.23	3.60
Phenol	16.65	62.88	49.38	57.34
Oxygenated compounds	26.06	9.08	20.29	27.70
Total selectivity	100.00	100.00	100.00	100.00

Reaction conditions: Catalysts = 3Ga/HBeta, 3Ga/Silica, 3Ga/HZSM-5, and HZSM-5, W/F = 6.3 h, Reaction temperature = 450°C, Carrier gas = H_2, Pressure = 1 atm.

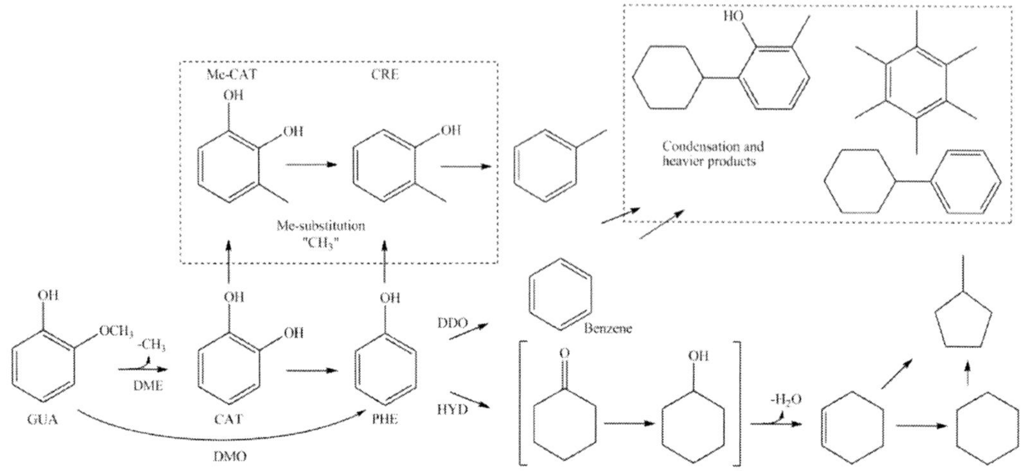

DME, demethylation; DMO, demethoxylation; DDO, direct deoxygenation; GUA, guaiacol; CAT, catechol; PHE, phenol; Me-CAT, methyl-catechol; Me-PHE, methyl-phenol; CRE, cresol.

Scheme 2. General reaction scheme for guaiacol conversion [15].

In the case of unsupported CoMoS catalyst, the major deoxygenated product was benzene. Supported CoMoS catalyst also produced some benzene, but in a lesser extent. This indicates that the CoMoS phase favors direct cleavage of aromatic carbon-heteroatom bond at the expense of hydrogenation route, similarly to the hydrodesulphurization reactions. For supported catalysts, the distribution of deoxygenated methyl-substituted products logically issued from the methyl-phenols, was different depending on the catalysts. According to disproportionation activity, CoMoS/γ-Al_2O_3 gave mostly aromatic methylated compounds like toluene whereas, over MoS_2/γ-Al_2O_3 catalyst, the main O-free product was methyl-cyclohexene.

Deoxygenation of phenolic compounds has become major concern for upgrading pyrolysis product due to a large consumption of H_2 for hydrotreating approaches. To avoid large consumption of H_2, new approaches involve weakening of C-O bonds without ring saturation (Hydrogenation reaction). All these features provide an opportunity to potentially design catalysts at a molecular level with specificity and activity for catalytic system.

Hydrodeoxygenation of guaiacol was also tested on a series of novel hydroprocessing catalysts – transition metal phosphides which included Ni_2P/SiO_2, Fe_2P/SiO_2, MoP/SiO_2, Co_2P/SiO_2 and WP/SiO_2 at 300°C and atmospheric pressure [16]. The benzene selectivity for hydrodeoxygenation of guaiacol follows the order: Ni_2P/SiO_2 > Co_2P/SiO_2 >>> WP/SiO_2 (Table 4).

The products for catalytic conversion of guaiacol are phenol (product from demethoxylation), benzene (product from demethoxylation and/or direct deoxygenation), methoxybenzene (product from direct deoxygenation), with no catechol (product from demethylation) formed at higher contact time (20.2 min, space velocity = 1.4 h^{-1}) (Table 4). At lower contact time (0.339 min, space velocity = 59 h^{-1}) catechol is the major products for Co_2P/SiO_2 and WP/SiO_2. No catechol was observed for hydrodeoxygenation of guaiacol with Ni_2P/SiO_2 even at low contact time.

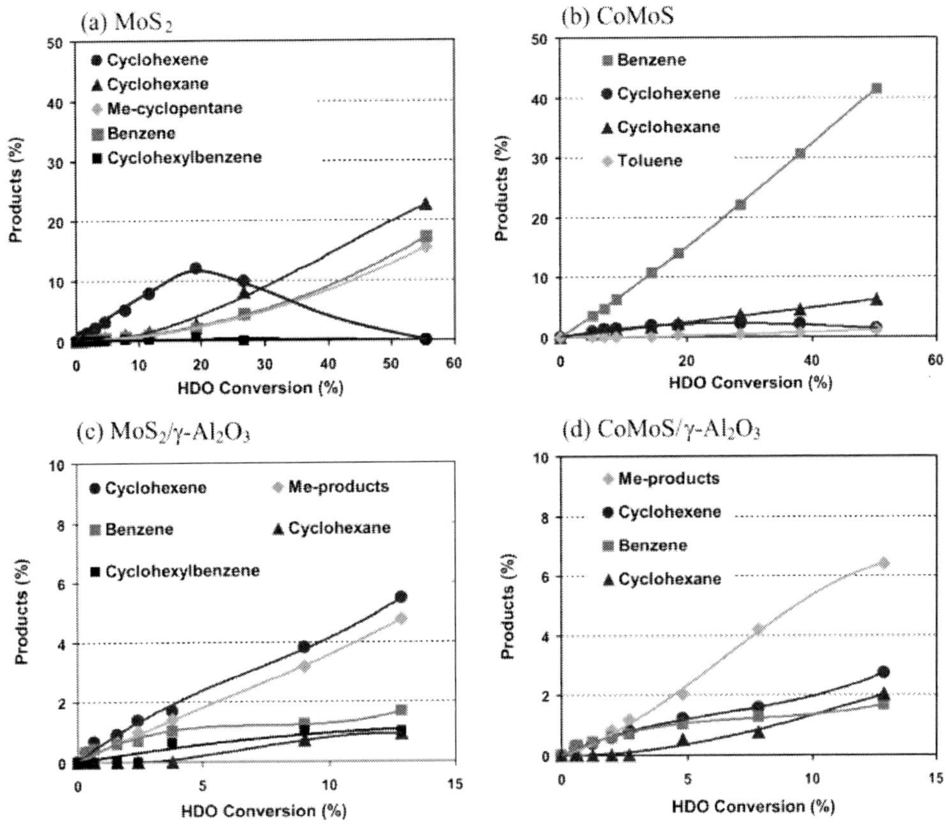

Figure 1. O-free products distribution versus total deoxygenation during guaiacol conversion at 300°C. (a) with MoS_2, (b) with CoMoS, (c) with $MoS_2/\gamma\text{-}Al_2O_3$, and (d) with $CoMoS/\gamma\text{-}Al_2O_3$ [15].

Table 4. Effect of catalyst type and contact time on conversion of guaiacol at 300°C [16]

Catalyst	Contact time (min)	Space velocity (h^{-1})	Conversion (%)	Product distribution (%)					
				PHE	BZ	MB	CAT	CRE	C_3-C_5
Ni_2P/SiO_2	20.2	1.4	80	30	60	10	0	0	0
	0.339	59	19	28	4	38	0	30	0
Co_2P/SiO_2	20.2	1.4	70	32	52	1	0	0	15
	0.339	59	35	1	0	0	99	0	0
WP/SiO_2	20.2	1.4	60	100	0	0	0	0	0
	0.339	59	12	12	0	0	88	0	0

PHE = phenol, BZ = Benzene, MB = Methoxybenzene, CAT = Catechol, CRE = Cresol, C_3-C_5 = C_3-C_5 hydrocarbons.

The commercial 5% Pd/Al_2O_3 catalyst was compared to such metal phosphide catalyst. It shows a higher activity as compared to the metal phosphides at lower contact time, but the major product is undesired catechol. The commercial hydrotreating catalyst $CoMoS/Al_2O_3$ deactivated quickly and showed little activity for the hydrodeoxygenation of guaiacol at these conditions. These results indicate that transition metal phosphides are promising catalysts for the treatment of bio-derived feedstocks.

From the above literature, it can be concluded that phenolics can also be coupled to larger oxygenate species over bifunctional catalysts (i.e. Ga/HBEA). Together with H-transfer, active metal readily promotes hydrogenolysis of larger precursors since cleavage of the C-O bridging in the bulky isolated rings can be easily facilitated, as compared to that in the single ring aromatics. This approach involves weakening of C-O bonds without ring saturation. The other approach, the oxygenated substitution group in phenolics can be readily removed via direct deoxygenation over transition metal compound (i.e. metal sulphide and metal phosphide). The latter approach involves direct hydrogenolysis of C-O bonds without ring saturation.

The Production of BTEX from Aromatic Ether (Anisole)

The catalytic conversion of anisole (methoxybenzene) has been investigated over a bifunctional Pt/HBeta catalyst at 400°C and atmospheric pressure [17]. The product distribution obtained on the bifunctional catalyst was compared with those obtained on monofunctional catalysts (HBeta and Pt/SiO_2). This comparison indicates that the acidic function (HBeta) catalyzes the methyl transfer reaction (transalkylation) from methoxyl to the phenolic ring, yielding phenol, cresols, and xylenols as the major products.

The metal function catalyzes demethylation, hydrodeoxygenation, and hydrogenation in sequence, resulting in phenol, benzene, and cyclohexane. On the bifunctional catalyst, both methyl transfer and hydrodeoxygenation are achieved at significantly higher rates as compared to that over the monofunctional catalysts. This is crucial for the formation of benzene, toluene, and xylenes with lower hydrogen consumption. Moreover a significant reduction in carbon losses was observed, in comparison to the system with the metal function alone.

Table 5. Major product distributions for complete conversion of anisole over HBeta and complete deoxygenation of anisole over Pt/SiO_2 and Pt/HBeta catalysts [17]

Catalyst	HBeta	1% Pt/SiO_2	1% Pt/HBeta	0.2% Pt/HBeta
W/F (h)	0.5	4	0.33	1.5
Major product yield (%)				
Phenol	39.5	0.2	0.1	0.7
Cresols	33.5	0	0.2	0.6
Xylenols	24.2	0	0	0.3
C_{1-2}	0.2	15.4[a]	6.0	4.2
C_3	0.2	0	0.4	0.7
C_{4-9}	0.2	14.1	1.6	1.5
Benzene	0.1	69.2	51.2	40.2
Toluene	0.1	0.8	27.6	31.8
Xylene	0.3	0	10.6	14.4
C_{9+} aromatics	0.3	0	1.7	3.4

[a] CH_4 is the only product for C_{1-2} aliphatic hydrocarbons.

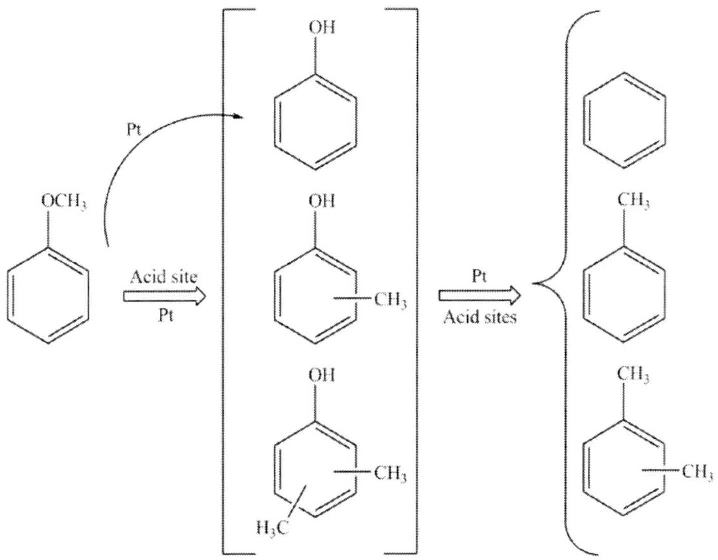

Scheme 3. A major reaction pathway of deoxygenation of anisole over Pt/HBeta catalyst [17].

This result suggests that by adjusting the metal/acid balance, one could control product distribution and improve aromatic yield without excess hydrogen consumption. In addition, the incorporation of metal into acidic zeolites improves the stability of the catalyst with a moderate reduction in coke formation.

From this approach, the acidic oxides of zeolite plays important role for the reactions involving disoproportionation that retain methyl group in the products mixture. This leads to the formation of xylene despite the pyrolysis feed contains carbon in the molecule less than C8 aromatics.

The Production of BTEX Aromatics from Aromatic Aldehyde (Benzaldehyde)

Not only for phenolic compounds, aromatic aldehydes have been investigated for aromatic production. The deoxygenation of benzaldehyde, a model compound represent aromatic aldehyde from pyrolysed products, has been tested over gallium-modified HZSM-5 catalysts at 450-500°C, atmospheric pressure [18].

From Table 6, in the absence of H_2, Ga/HZSM-5 catalyzes decarbonylation of benzaldehyde resulting in benzene and CO. The active sites for this reaction are the strong Brønsted acid sites. In the presence of H_2, the main product is toluene. It is believed that Ga cationic species (Ga^+/GaH_2^+) generated during H_2 reduction can promote the hydrogenation/hydrogenolysis reactions that give toluene and water (Scheme 4).

At lower W/F (< 100 g.h/mol), the increase in benzene and toluene yields with W/F suggests that they are formed in parallel as primary products,. The small amount of methane may be the result of hydrodealkylation of toluene. However, heavier secondary products such as xylenes and condensation oxygenated compounds are also observed at higher W/F. Xylenes are possibly derived from the toluene disproportionation, while the oxygenates can be generated from the secondary reaction of benzaldehyde with the decarbonylated / hydrogenolyzed species. It is indicating that reaction conditions favoring the formation of xylenes products are also important.

The deoxygenation of benzaldehyde to benzene and toluene was also investigated on basic CsNaX, and NaX zeolite catalysts [19]. It was observed that as-prepared CsNaX, containing Cs in excess, displays high activity for direct decarbonylation of benzaldehyde to benzene (Figure 2).

However, in parallel with the decarbonylation reaction, condensation of the products on surface also occurs. Decomposition of the surface condensation products results in further evolution of benzene and toluene. It is observed that gas-phase H_2 can play an important role by reducing the residence time of surface intermediates, thus decreasing the amount of condensation products that accumulate and lead to catalyst deactivation. Hydrogen transfer to the condensed products accelerates the decomposition of these condensation compounds primarily into toluene. NaX catalyst and washed CsNaX do not exhibit a high initial activity for direct decarbonylation, but rather proceed via formation of surface condensation products which subsequently decompose to benzene and toluene. The residual acidity present in NaX catalysts causes a faster deactivation for this catalyst than for those containing Cs (Scheme 5).

Table 6. Product distribution for conversion of benzaldehyde [18]

Type of catalysts	Temperature (°C)	Carrier gas	% Conversion	Product distribution (% yield)		
				Benzene	toluene	Methane
HZSM-5	450	He	56.32	56.32	0.00	0.00
		H_2	54.23	54.23	0.00	0.00
3Ga/H ZSM-5	450	He	55.80	55.80	0.00	0.00
		H_2	58.20	19.95	36.40	1.85
3Ga/H ZSM-5	500	He	69.07	69.07	0.00	0.00
		H_2	70.22	20.42	43.71	6.09

Reaction conditions: Catalysts = HZSM-5 and 3Ga/HZSM-5, W/F = 100 g·h/mol, Reaction temperature = 450 and 500°C, Carrier gas = He and H_2, Pressure = 1 atm.

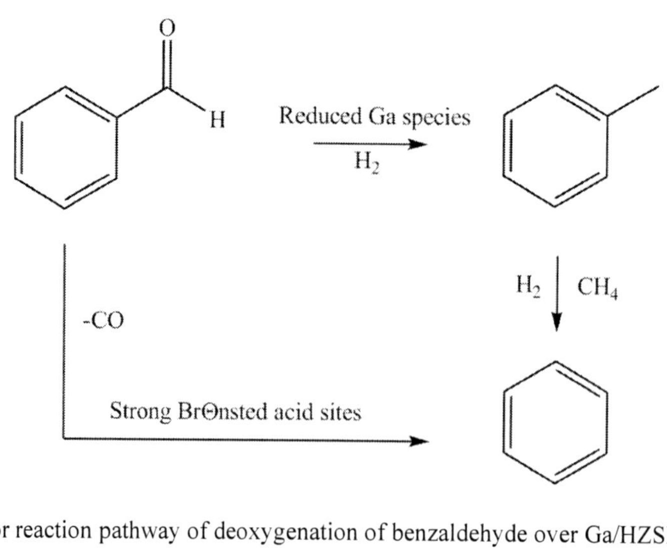

Scheme 4. A major reaction pathway of deoxygenation of benzaldehyde over Ga/HZSM-5 catalyst [18].

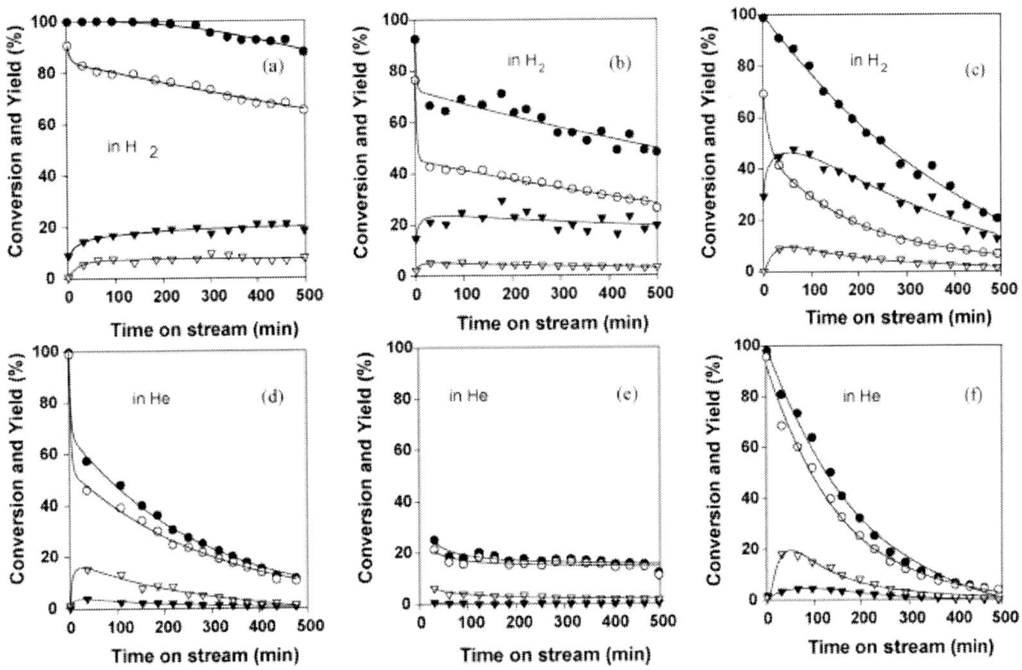

Figure 2. Conversion and product distribution from the flow reaction of benzaldehyde over Cs and NaX catalysts at 475°C at W/F = 942 g.h/mol using different carrier gases; (a and d) CsNaX, (b and e) washed CsNaX, and (c and f) NaX. Conversion (●), benzene (○), toluene (▼), and 2-ring compounds (∇) [19].

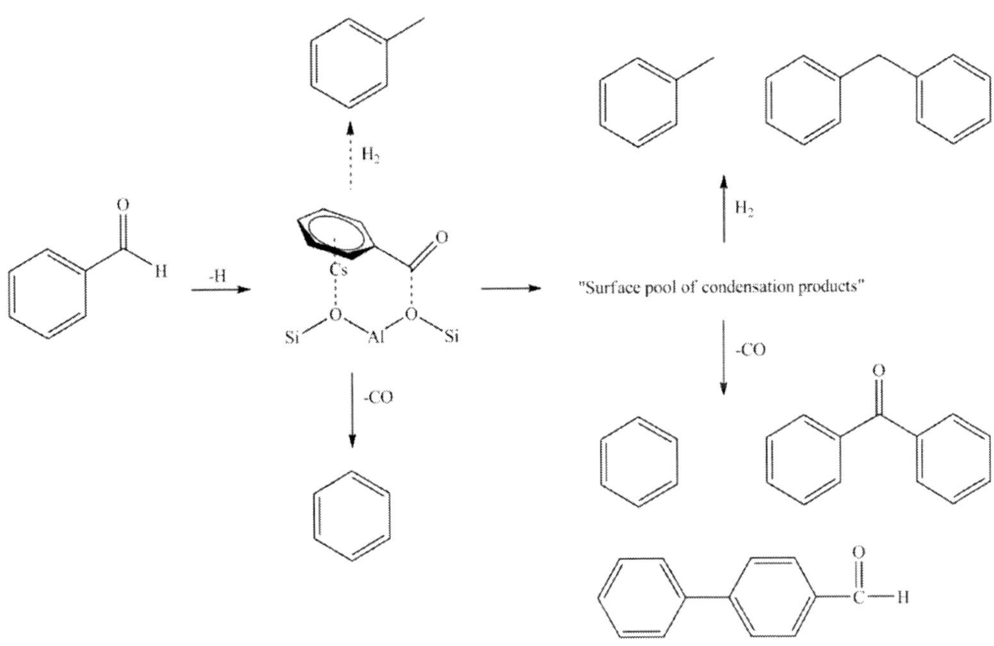

Scheme 5. A major reaction pathway of deoxygenation of benzaldehyde over CsNaX zeolite [19].

From the above results, it is confirmed that the surface pool intermediates can be formed by acidic function. Such surface pool intermediates can be removed by hydrogenolysis activity to lighter aromatic products (benzene and toluene), improving the catalyst stability. Therefore, the balance between these competing reaction pathways determines whether the catalyst shows stability or deactivation. However, lack of disproportionation activity of the basic site eliminates the formation of xylenes and higher aromatics.

Conversion of Refined Biomass

Considerable efforts for utilizing renewable resources for xylene production have also been made from refined biomass such as ethanol, fatty acid and glucose. This approach involves dehydration, decarbonylation and/or decarboxylation to primarily produce olefinic or paraffinic hydrocarbons that can be further converted to aromatic via classical acid catalysed processes. Shape selective feature of the acid catalyst (zeolites) always plays marked role for selective formation of xylenes.

The Production of BTEX Aromatics from Ethanol

Ethanol has been produced mainly from corn in North America and China and from sugarcane in Brazil. However, since corn is a major food source, its use as a fuel raw material has been criticized as it has led to a dramatic increase in the price of corn. Therefore, waste biomass such as corn stover, waste wood and waste food, as cheap raw material for ethanol production, are much more attractive than corn.

For the production of aromatics, ethanol must be converted in two-steps: catalytic dehydration of ethanol to ethylene and subsequent conversion of ethylene to hydrocarbons, which proceed via classical carbocation mechanisms with concurrent hydrogen transfer. It is noting that the shape selective catalyst i.e. ZSM-5 zeolite, played a vital role in determining the nature of the product, especially its framework topology, and Si/Al, which is related to the acidity of the catalyst, [20-22].

The high amount of gas obtained on low acidity catalyst can be recognized from the reaction pathways for ethanol conversion, where the formation of aromatics comprises the sequential steps of ethanol dehydration to ethylene, oligomerization, cracking and/or cyclization and hydrogen transfer reactions (Scheme 6). Thus the formation of lighter hydrocarbons (C_2-C_4) is considered to occur from initial dehydration of ethanol, dimerization of ethylene and its cracking [23].

The ethanol conversion is conducted in a fixed bed down flow reactor. The feed was introduced in a continuous flow using Eldex plunger type feed pump. While using N_2 as carrier gas the reaction is conducted at 450-500°C, atmospheric pressure and 2.0-3.0 h^{-1} WHSV. The product yield was strongly dependent on SiO_2/Al_2O_3 ratios as well as on the reaction conditions [24].

The products obtained at this reaction were gases (30-65 wt.%) and liquids (35-70 wt.%, Aromatic ~ 40-50% of liquid yield) (Table 7). It is known that higher hydrocarbons can be readily cracked into lower hydrocarbons at high temperature. While, olefins are readily converted to higher molecular weight hydrocarbons which undergo isomerization to high quality gasoline over HZSM-5 at low reaction temperature (< 400°C). However, an effort to convert gasoline to more valuable petrochemical products has focused on improving the conversion of gasoline to aromatic hydrocarbons. It is found that aromatic yield first increase with rise in temperature (350-425°C). This is because the higher temperature is favored for hydrogen transfer and the reaction of higher hydrocarbon products shifts toward the aromatic fraction. At > 450°C, the aromatic product is decreased due to inhibition of higher hydrocarbon formation. This is due to the fact that ethylene oligomerization is an important step for the production of aromatic. Therefore, a balance between oligomerization and hydrogen transfer is essential for obtaining high yield of aromatics.

The incorporation of metal, namely Ag, Ga, and Fe into HZSM-5 zeolite provides the increase in aromatic yield [25-27].

In Table 8, higher BTX selectivity is readily obtained from AgHZSM-5(11), as compared with HZSM-5(11) which also well agrees with the other reports [28-29]. However, it is known that different catalytic activities and the reaction pathway, and hence product selectivity are pronounced when different metal species (metal cations, cationic clusters, metallic clusters or oxides formed) are introduced.

While, alkaline earth metal doped HZSM-5 zeolites exhibited the highest C_3H_6 yield (32 %C yield) and a high catalytic stability at the reaction condition of 500°C and W/F value of 0.03 g_{cat}/ml/min. [30].

Zeolite modified phosphorous provides the high selectivity of propylene at low reaction temperature. However, the main components of liquid products are aromatics, mostly toluene and xylene (Table 9). The fraction of xylene in liquid hydrocarbons is far more pronounced over the P/HZSM-5 catalyst, as compared to that of the parent HZSM-5 at high reaction temperature [31]. It is worth noting that the presence of water in feed can inhibit ethylene oligomerization [25] leading to the lower yield of aromatics (Table 10).

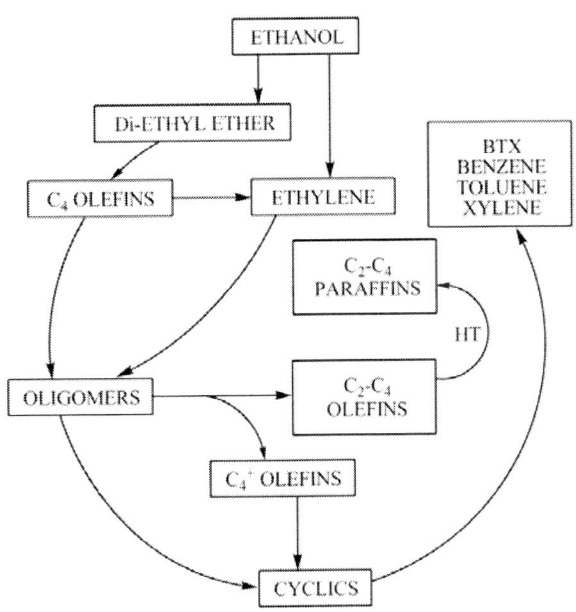

Scheme 6. Reaction pathways for ethanol to aromatics [24].

Table 7. Product distribution for conversion of ethanol over various zeolite catalysts [24]

Catalysts	**NZ**	**Z**	**ZA**	**NZ**	**Z**	**ZA**
Temperature (°C)		450			500	
Gas analysis (wt.%)						
Gas composition	31.5	37.6	57.6	29.9	48.7	64.8
C_1	0.2	0.5	0.4	0.4	0.2	2.7
$C_2^=$	7.5	4.7	5.2	2.7	6.5	4.8
C_2	0.6	0.9	2.2	1.3	1.2	1.9
$C_3^=$	5.4	4.4	7.4	3.4	12.6	8.0
C_3	3.0	10.6	19.7	10.5	13.8	33.5
iC_4	5.7	9.6	12.5	6.2	2.5	7.9
C_4	2.3	4.4	5.1	2.6	0.8	4.0
$C_4^=$	3.3	2.5	3.3	1.5	0.5	2.0
iC_5	3.5	-	1.8	1.3	7.0	-
nC_5	-	-	-	-	1.8	-
$C_5^=$	-	-	-	-	1.8	-
Liquid analysis (wt.%)						
Liquid composition	68.5	62.4	42.4	70.1	51.3	35.2
Paraffins	1.0	0.3	0.4	1.7	1.1	0.0
C_5	0.1	0.1	0.1	0.1	-	0.0
C_6	0.1	0.1	0.1	0.1	0.0	0.0
C_7	0.4	0.1	0.1	0.4	0.2	-
C_8	0.4	-	-	0.8	0.3	-
C_9	-	-	-	0.1	0.6	-
C_{10}	-	-	0.1	0.2	-	-
i-Paraffins	11.4	13.0	4.1	9.5	8.5	2.9
C_5	0.4	0.7	0.1	0.2	0.1	0.1
C_6	0.6	0.3	0.2	0.5	0.1	0.1
C_7	0.8	0.4	0.2	0.7	0.2	0.1
C_8	1.0	0.4	0.1	1.0	0.2	0.1
C_9	0.7	0.1	2.7	0.1	-	1.7
C_{10}	7.9	11.1	0.8	7.0	7.9	0.8
Aromatics	38.6	38.7	35.3	50.6	36.3	31.8
C_6	0.5	0.9	1.0	2.5	1.1	1.8
C_7	7.9	11.4	10.1	17.0	11.7	11.5
C_8	24.0	26.1	16.9	27.5	22.3	14.7
C_9	5.8	-	7.3	2.9	0.5	3.8
C_{10}	0.4	0.3	-	0.7	0.7	-
Naphthenes	3.5	1.6	1.0	1.8	0.6	0.2
C_5	0.1	0.1	0.1	0.1	0.0	0.0
C_6	0.6	0.4	0.3	0.4	0.1	0.1
C_7	0.8	0.6	0.3	0.5	0.2	0.1
C_8	1.5	0.5	0.2	0.6	0.3	-
C_9	0.1	-	-	-	-	-
C_{10}	0.4	-	0.1	0.2	-	-

Catalysts	NZ	Z	ZA	NZ	Z	ZA
n-Olefins	0.4	0.1	0.1	0.9	0.0	0.1
C_5	0.0	-	0.0	0.0	0.0	0.0
C_6	0.3	-	0.0	0.1	-	-
C_7	0.1	0.1	0.0	0.8	0.0	0.0
C_9	-	-	0.1	-	-	0.1
Iso-olefins	11.7	7.7	1.0	5.2	4.1	0.1
C_5	0.4	0.1	0.1	0.2	0.0	0.1
C_6	0.3	0.1	0.1	0.1	0.0	0.0
C_7	0.8	0.3	0.0	0.2	0.1	0.0
C_8	0.1	0.1	-	-	0.0	-
C_{10}	10.1	7.1	0.8	4.7	4.0	0.0
Di-olefins	0.1	0.0	-	0.1	0.0	-
C_5	0.1	-	0.0	0.0	-	-
C_7	-	-	0.0	0.1	0.0	-
Naphtheno-olefins	1.1	0.2	0.4	0.1	0.2	0.0
C_5	0.3	-	0.0	-	0.1	-
C_6	0.8	0.1	0.1	0.1	-	0.0
C_7	-	0.1	0.3	-	0.1	-
Others	0.7	0.8	0.1	0.2	0.5	0.1
Total	100.0	100.0	100.0	100.0	100.0	100.0
RON	91.0	87.1	105.5	95.4	90.5	108.4

NZ; nano crystalline ZSM-5, Z; micro crystalline ZSM-5, ZA; micro crystalline ZSM-5 with low acidity.

It is clearly because the competitive adsorption of water can reduced ethylene oligomerization. For the ethanol feed, the majority of the containing water must be eliminated. Most of the water is removed by distillation, but the purity is limited to 95-96% due to the formation of a low-boiling water-ethanol azeotrope with maximum 95.6% w/w (96.5% v/v) ethanol and 4.4% w/w (3.5% v/v) water. The water fraction is typically removed in further treatment such as pervaporation in order to reduce the competitive adsorption.

Table 8. Product distribution for conversion of ethanol over Ag modified HZSM-5 catalysts[a] [26]

Catalyst	Product distribution (% mol)[b]					
	C2	C3	C4	C5-C8	BTX[c]	C9+
HZSM-5(11)	13.3	29.4	27.7	1.9	17.8	9.8
AgHZSM-5(11)	19.8	15.7	16.7	1.9	27.4	18.5
HZSM-5(28)	21.6	25.0	23.9	4.3	13.0	12.2
AgHZSM-5(28)	96.2	0.0	1.3	0.1	0.7	1.7
AgHZSM-5(28)/H_2	36.2	23.0	24.6	4.0	6.1	6.2

[a] Reaction conditions : temperature = 425°C, total pressure = 1 atm, carrier gas = N_2 or H_2, W/F = 6.75 g·h/mol.[b] Product distribution at 1 h on stream.[c] BTX = benzene, toluene, ethylbenzene and xylenes.

Table 9. Product distributions for conversion of ethanol over P modified HZSM-5 catalysts [31]

Temperature (°C)	Conversion (%)	Selectivity (%)				
		$C_2^=$	$C_3^=$	$C_4^=$	C_1-C_4	C_6-C_9 aromatics
270	81.5	95.5	0.4	0.2	0.3	3.6
300	99.4	77.7	3.5	4.1	3.4	11.3
330	99.9	54.3	9.6	9.2	11.8	15.1
360	100	48.5	10.4	9.6	12.1	19.4
390	100	28.9	17.5	15.9	14.6	23.1
420	100	22.6	28.0	13.3	18.5	27.6
450	100	34.6	18.9	10.1	15.9	24.3
470	100	31.3	14.3	10.9	19.9	23.6

Reaction condition: WHSV = 3.16 h^{-1}, 0.3 g catalyst, pure ethanol, P = 1 atm.

Table 10. Effect of water co-feeding over Ga modified HZSM-5 on the conversion of ethanol[a] [25]

Catalyst	Feed	Product distribution (% mol)[b]				
		C_2	C_3	C_4	C_5^+	BTX
HZSM-5	95 wt.% Ethanol	23.61	37.47	26.19	8.42	4.31
	Absolute ethanol	17.44	36.96	28.13	9.91	7.55
	Ethylene	13.86	39.93	29.50	9.18	7.53
[Ga] HZSM-5	95 wt.% Ethanol	17.83	34.22	28.76	11.03	8.16
	Absolute ethanol	18.16	34.52	29.19	11.53	6.60
	Ethylene	34.86	22.14	23.88	8.92	10.19

[a] Pressure = 1 atm, temperature = 425°C, WHSV = 2.88 h^{-1}, [b] Product distributions were averaged over 1-3 h on stream.

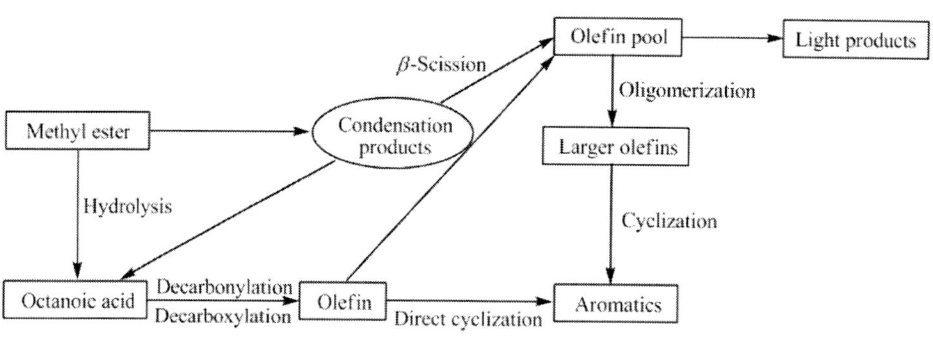

Scheme 7. Proposed reaction scheme for the conversion of methyl octanoate over the HZSM-5 catalyst [32].

Interestingly, in the case of [Ga]HZSM-5, selectivity of C_4^+ hydrocarbons and BTX increases with rise in water content (in the case of 95 wt.% ethanol). This is because the presence of water can produce additional acid sites in [Ga]HZSM-5 which can remain at the reaction temperature (425°C). The enhanced acidity of [Ga]HZSM-5 can promote ethylene oligomerization and the subsequent aromatization of higher hydrocarbons. From the above results, it can be concluded that the xylene production can be readily regulated by temperature and the catalytic behaviour of the additional active species.

The Production of BTEX Aromatics from Ester and Fatty Acid

Fatty acid methyl esters (FAME) are a type of fatty acid ester that can be produced by an alkali- or acid-catalyzed reaction of fats or fatty acids and methanol. The biodiesel is originally FAMEs, usually obtained from vegetable oils by transesterification. A fatty acid is a carboxylic acid with a long aliphatic tail that is either saturated or unsaturated. Most naturally occurring fatty acids have a chain of an even number of carbon atoms, from 4 to 28. Fatty acids are usually derived from triglycerides or phospholipids. When they are not attached to triglyceride, they are known as "free" fatty acids. Both fatty acid methyl esters and fatty acids are important sources of fuel and petrochemical feedstocks.

The conversion of the fatty acid methyl ester, namely methyl octanoate, a model compound representative of bio oil, was studied over HZSM-5 catalyst [32]. The proposed reaction to produce aromatics, follows two parallel paths that start with self-condensation and hydrolysis, respectively. The condensation products can undergo subsequent reactions to form both non-aromatic and aromatic secondary products. For the second path, the water required for the hydrolysis of ester can be supplied from the dehydration of condensation products and the subsequent decomposition of dimethyl ether. The octanoic acid formed by this route can be decarboxylated / decarbonylated forming olefins, which are incorporated to the common olefin pool, eventually becoming aromatics (Scheme 7).

Table 11. Product distribution for conversion of methyl octanoate over HZSM-5 catalyst[32]

Reaction temperature (oC)	500	500	400
W/F ($g_{cat}h/mol_{MEO}$)	6.4	0.8	6.4
Conversion	99.7	56.5	48.6
Product selectivity (wt.%)			
C_1	6.7	0.5	0.6
C_2	34.3	13.7	3.3
C_3	15.3	1.4	2.2
C_4	5.3	0.2	0.2
C_5	5.8	2.7	0.6
C_6	3.7	1.8	1.0
C_7	2.2	6.5	7.3
Benzene	1.6	0.6	0.6
Toluene	6.2	0.4	0.4
C_8 aromatics			
Ethylbenzene	1.0	0.2	0.1
m-Xylene	8.1	1.0	1.2
p-Xylene	1.8	0.1	0.1
o-Xylene	0.2	0.3	0.1
C_9^+ aromatics	2.9	1.6	1.6
Octanal	1.1	0.1	1.0
Octanoic acid	0.4	22.7	50.4
Coupling products			
Tetradecene	2.6	2.9	16.0
8-Pentadecanone	0.6	0.3	0.5
Coupling ester	0.2	43.2	12.8

Reaction conditions: P = 1 atm; H_2:feed = 6:1 molar ratio; time on stream = 20 min.

The conversions of methyl octanoate are conducted at 400 and 500°C reaction temperature, atmospheric pressure and 0.8 and 6.4 W/F ($g_{cat.}$ h/mol_{MEO}).

Table 11 shows that at 500°C, C_2 and C_3 hydrocarbons are the dominant products (~50 wt.%). In addition, a significant amount of aromatics (~22 wt.%) were observed. By contrast, at 400°C at the same W/F and time on stream, the conversion is much lower and octanoic acid and condensation products (namely 8-pentadecanone, tetradecene, and condensation ester) were the major products. Much lower yields of total aromatics were obtained at this temperature.

In the case of using fatty acid methyl ester from bio-oil (C_{12}-C_{20}) as a feed, it is expected that a slightly difference between the selectivity to total aromatics can be observed from methyl octanoate as a function of W/F. This is because aromatics can be formed through a series of reactions, namely cracking, oligomerization, and cyclization. Therefore, the product selectivity can be regulated by the restricted pore size of ZSM-5 zeolite.

In the case of methyl octanoate conversion over a basic zeolite catalyst, CsNaX [33], a remarkable difference between the product selectivity obtained from HZSM-5 and CsNaX can be observed.

The major product is C_7 hydrocarbons (i.e. heptenes), which is the expected product from direct decarbonylation of methyl octanoate over CsNaX (425°C, 1 atm, W/F = 198 g h/mol). In addition, significant amounts of hexane, a product of deacetalation, are obtained. In contrast with the product distribution obtained from CsNaX, it was observed that over NaX catalyst hexenes, heptenes, and multisubstituted aromatics were the dominant products at the beginning of the reaction (Figure 3). Then, as the catalyst deactivated, the aromatic products tended to disappear and only hexenes and heptenes were obtained as main products at longer time on stream. It is evidenced the role of acid sites which can undergo classical carbocation mechanisms with concurrent hydrogen transfer to form aromatics.

Table 12. Product distribution for conversion of methyl octanoate over CsNaX catalyst[33]

	W/F (g h/mol)		
	99	**198**	**396**
Conversion	24.0	37.2	62.3
Yield (%)			
Hexenes	4.5	8.8	19.3
n-Hexane	0.2	0.1	0.3
Heptenes	8.9	14.6	26.1
n-Heptane	0.1	0.2	0.8
Octenes	3.0	6.1	6.7
n-Octane	0.4	0.8	1.0
Octanal	0.6	0.5	0.4
Oxygenates	1.9	2.0	3.3
Octanoic acid	0.8	1.1	2.2
Tetradecene	0.3	0.3	0.6
Pentadecanone	2.0	1.5	0.9
Coupling ester	1.2	1.1	0.8

Reaction conditions; 425°C, 1 atm, 25 mL/min of He.

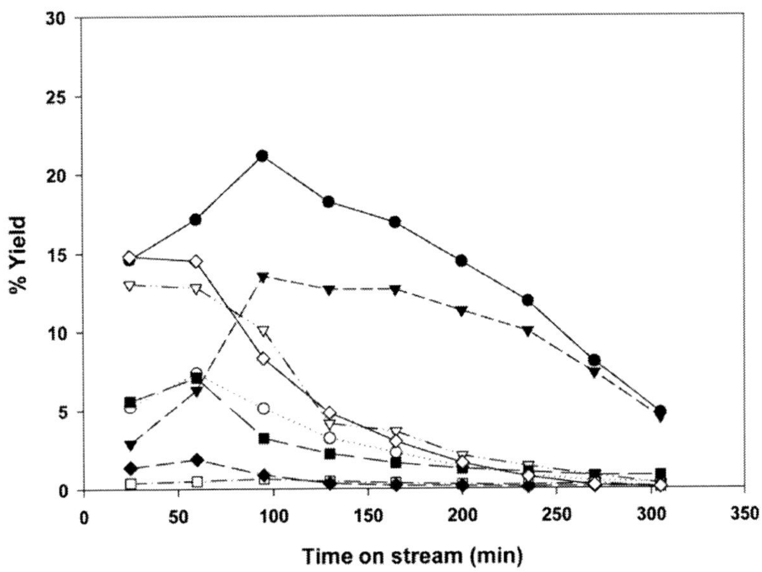

Figure 3. Product distribution from the reaction of 10 wt.% methyl octanoate in methanol over NaX zeolite catalyst as a function of time on stream [33]. Hexenes (●), heptanes (▼), multisubstituted aromatics (◊), heptanes (∇), octenes (■), hexane (○), octane (□), and coupling hydrocarbons (♦). Reaction conditions: 425°C, 1 atm, W/F = 198 g.h/mol, 25 mL/min of He.

The Production of BTEX Aromatics from Glucose Derivatives

The exploitation of ligno-cellulosic material in generation processes is crucial to avoid the competition with an increasing food demand and the social aspects related to this. Considerable efforts have been devoted up to now to the depolymerization of cellulose. In fact the tight packing of cellulose chains in microfibrils protects the β-1,4-glycosidic linkages between the glucose units, making the cellulose hard to attack and deconstruct over mineral acids, such as H_2SO_4 [34], HCl [35], HF [36], enzymes [37], solid acid catalysts [38]. However, once hydrolyzed, isomerization of glucose to fructose can be performed using biological (i.e., enzymes) or thermochemical catalysts including base catalysts [39] or Lewis acid heterogeneous catalysts such as Sn-Beta.[40]. Fructose can then be converted to hydroxymethylfurfural [41], which can be hydrodeoxygenated to produce dimethylfuran [42] as seen in an overall process designed to produce ligno-cellulosic material (Scheme 8).

To produce p-xylene, dimethylfuran must then be converted in two-steps: Diels-Alder cycloaddition of ethylene followed by dehydration [43] (Scheme 9).

The initial step in the conversion of dimethylfuran to *p*-xylene is thought to occur by the Diels-Alder reaction with ethylene as the dienophile. It is known that furans undergo [4+2]-cycloaddition with a number of dienophiles, including alkynes and alkenes, but the reaction environment and functionality of the diene/dienophile leads to significant variation in activity and stereochemistry. Employing ethylene as the dienophile in Diels-Alder processes is inherently difficult; ethylene lacks an electron-withdrawing group that promotes the reaction by closing the HOMO-LUMO gap of the interacting molecules. For a large number of substituted furans, [4+2]-cycloaddition with an alkene results in the formation of an oxabicyclic cycloadduct. This step has been catalyzed by a number of Lewis acids, including $ZnCl_2$, ZnI_2, and Et_2AlCl. Similarly, cycloaddition of other dienes/dienophiles has been catalyzed by zeolites such as Y, beta and ZSM-5, some of which were exchanged with cations such as sodium or copper.

The dimethylfuran cycloaddition has been investigated over HY zeolite (Si/Al = 30) and 900 psig C_2H_4 with pure dimethylfuran at 200 to 300°C [43]. Increase in reaction temperature exhibits selectivity to *p*-xylene for high conversion of dimethylfuran, ($X_{DMF} > 90\%$) (Figure 4). Yield of *p*-xylene further increases by the addition of *n*-heptane solvent (25 vol % DMF/75 vol % *n*-heptane) at 300°C.

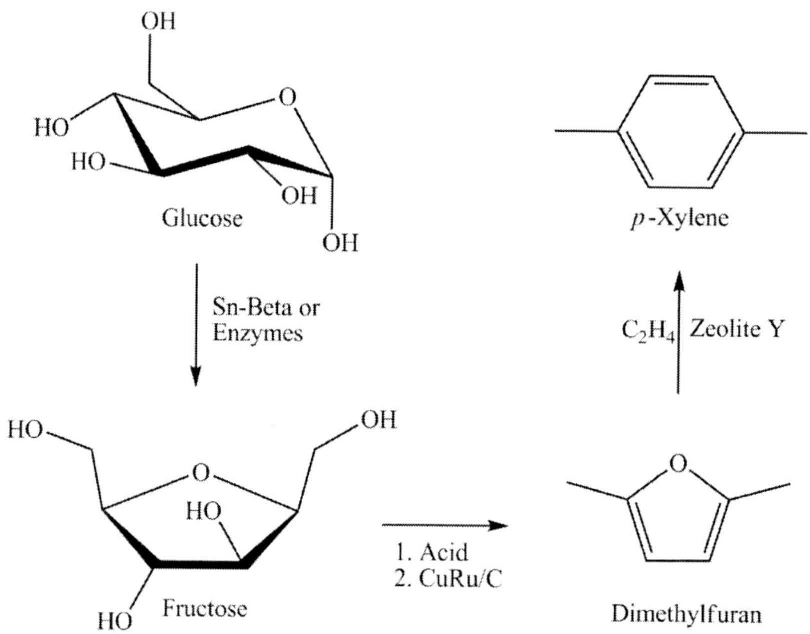

Scheme 8. Proposed reaction scheme for the conversion of glucose [43].

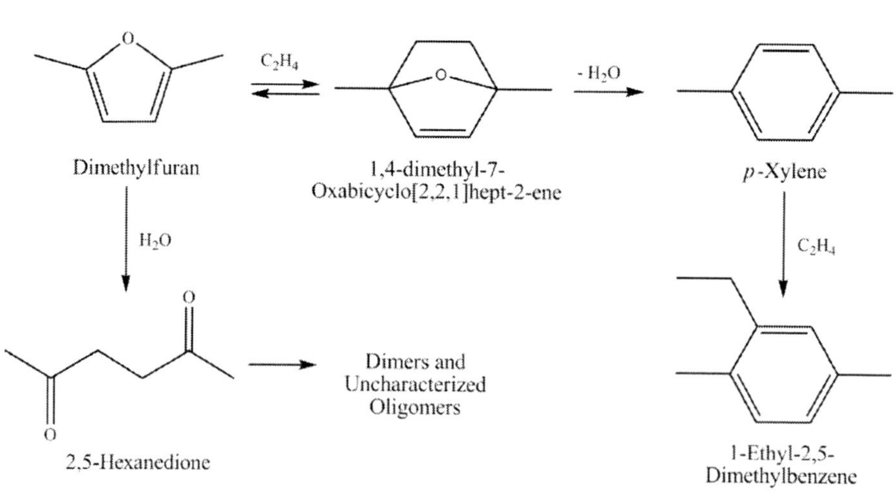

Scheme 9. Proposed reaction scheme for the production of *p*-xylene from glucose derivative [42].

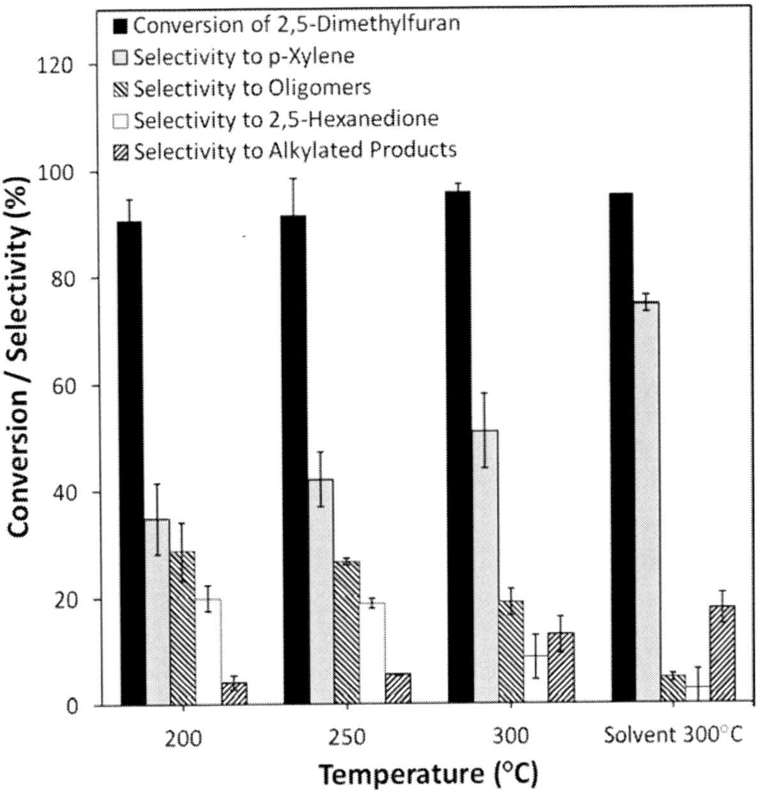

Figure 4. Demonstration of dimethylfuran cycloaddition at high conversion. Increase in reaction temperature from 200 to 300°C with HY zeolite Si/Al =30 and 900 psig C_2H_4 with pure dimethylfuran exhibits increasing selectivity to *p*-xylene for high conversion of dimethylfuran, $X_{DMF} > 90$ %. Yield of *p*-xylene further increases by the addition of *n*-heptane solvent (25 vol% DMF/75 vol% *n*-heptane) at 300°C [43].

CONCLUSION

Whereas most products of the chemical industry are based on petrochemical feedstock, considerable efforts have been made during the past few years to use renewable resources as industrial raw materials. A novel approach for the production of xylene from biomass-derived compounds was investigated. The production of aromatics from biomass-derived compounds can be achieved either by (i) pyrolysis bio-oil such as phenolic compounds, and benzaldehyde or (ii) refined biomass such as ethanol, ester, and fatty acid, glucose. From a chemical point of view, most biomass-derived compounds possess oxygenated reactive centers such as carbonyl, and hydroxyl functional group. Catalytic deoxygenation of such feedstocks has been attempted for aromatic production by using metal modified acid zeolite. It is well known that metal and acid sites are highly active for reactions involving hydrogen (hydrogenation / dehydrogenation and hydrogenolysis) and carbocation intermediates. Incorporation of metal into the catalysts is essential to promote hydrogenation, hydrogenolysis or hydrogen transfer. However, acidic function is necessarily required for the production of xylene via disproportionation, cyclization or aromatization. Zeolites always play key role on selectivity aspect for xylene production either from the use of H/C or biomass- derived feedstock.

REFERENCES

[1] Paolo, B; Pier, LB; Paolo, C; Giovanni, Z; Giuseppe, L; Enrico, M; Mario, P. *Zeolites*. 7, (1987), 418-422.
[2] Peng, W; Takayuki, K; Tatsuakim, Y. *Micropor. Mesopor. Mater.* 22, (1998), 343-356.
[3] Yong, JJ; Bin, Z; Le, X; Haihong, W; Honggen, P; Li, C; Yueming, L; Peng, W. *J. Catal.* 283, (2011), 168-177.
[4] Ikai, W; Tseng, CT; Sheng, TH. *Ind. Eng. Chem. Res.* 29, (1990), 2005-2012.
[5] Al-Khattaf, S; Tukur, NM; Al-Amer, A; Al-Mubaiyedh, UA. *Appl. Catal. A*, 305, (2006), 21-31.
[6] José, AD; Maria, CR; Sílvia, CLD; Ednéia, C; Fillipe, ACG. *Appl. Catal. A*, 328, (2007), 189-194.
[7] Andrea, K; Sulaiman, A; Muhammad, AA; Martina, B; Jiří, C. *Appl. Catal. A*, 377, (2010), 99-106.
[8] Wahab, A; Luqman, A; Rabindran, J; Sulaiman, A. *Chem. Eng. J.* 195-196, (2012), 276-288.
[9] Teresa, MP; Francisco, JL; Cristina, M; Susana, V; Avelino, C. *Appl. Catal. A*, 393, (2011), 257-268.
[10] Yan, Z; Wei, T; Hongyu, W; Anfeng, Z; Min, L; Guimin, L; Xiangsheng, W; Chunshan, S; Xinwen, G. *Catal. Today*, 160, (2011), 179-183.
[11] Nasiru, MT; Sulaiman, A. *Chem. Eng. J.* 166, (2011), 348-357.
[12] Toch, K; Thybaut, JW; Vandegehuchte, BD; Narasimhan, CSL; Domokos, L; Marin, GB. *Appl. Catal. A*, 425-426, (2012), 130-144.
[13] Balasamy, RJ; Odedairo, T; Al-Khattaf, S. *Appl. Catal. A*, 409-410, (2011), 223-233.
[14] Ausavasukhi, A; Yi, H; Anh, TT; Sooknoi, T; Resasco, DE. *J. Catal.* 290, (2012), 90-100.
[15] Van, NB; Dorothee, L; Pavel, A; Christophe, G. *Appl. Catal. B*, 101, (2011), 239-245.
[16] Zhaoa, HY; Lia, D; Buia, P; Oyama, ST. *Appl. Catal. A*, 391, (2011), 305-310.
[17] Zhu, X; Lobban, LL; Mallinson, RG; Resasco, DE. *J. Catal.* 281, (2011), 21-29.
[18] Ausavasukhi, A; Sooknoi, T; Resasco, DE. *J. Catal.* 268, (2009), 68-78.
[19] Peralta, MA; Sooknoi, T; Danuthai, T; Resasco, DE. *J. Mol. Catal. A: Chem.* 312, (2009), 78-86.
[20] Youming, N; Weiye, P; Guangxing, L. *J. Ind. Eng. Chem.* 16, (2010), 503-505.
[21] Song, Z; Takahashi, A; Mimura, N; Fujitani, T. *React. Kinet. Catal. Lett.* 131, (2009), 364-369.
[22] Murata, K; Inaba, M; Takahara, I. *J. Jpn. Petro. Inst.* 51, (2008), 234-239.
[23] Inaba, M; Murata, K; Saito, M; Takahara, I. Green Chem. 9, (2007), 638-646.
[24] Nagabhatla, V; Sandeep, KS; Jitendra, K; Peta, S; Devaki, N. *Fuel*, 95, (2012), 298-304.
[25] Ausavasukhi, A; Sooknoi, T. *Appl. Catal. A*, 361, (2009), 93-98.

[26] Ausavasukhi, A; Suwannaran, S; Limtrakul, J; Sooknoi, T. *Appl. Catal. A*, 345, (2008), 89-96.

[27] Valmir, C; Mauro, LB; Na′dia, RCFM. *Fuel*, 87, (2008), 1628-1636.

[28] Inoue, Y; Nakashiro, K; Ono, Y. *Microp. Mesop. Mater*. 4, (1994), 379-383.

[29] Ono, Y; Osako, M; Kim, GJ; Inoue, Y. *Stud. Surf. Sci. Catal*. 84, (1994), 1773-1780.

[30] Daisuke, G; Yasumitsu, H; Yoshiyasu, F; Atsushi, T; Tadahiro, F; Yasunori, O; Masahiro, S; Tsuneji, S. *Appl. Catal. A*, 383, (2010), 89-95.

[31] Jiangyin, L; Yancong, L. *J. Nat. Gas Chem*. 20, (2011), 162-166.

[32] Danuthai, T; Jongpatiwut, S; Rirksomboon, T; Osuwan, S; Resasco, DE. *Appl. Catal. A*, 361, (2009), 99-105.

[33] Sooknoi, T; Danuthai, T; Lobban, LL; Mallinson, RG; Resasco, DE. *J. Catal*. 258, (2008), 199-209.

[34] Wooley, R; Ruth, M; Sheehan, J; Ibsen, K. Lignocellulosic Biomass to Ethanol Process Design and Economics Utilizing Co-Current Dilute Acid Prehydrolysis and Enzymatic Hydrolysis Current and Futuristic Scenarios. NREL Report TP-580-26157, US Department of Energy, National Renewable Energy Laboratory, Golden, 1990, 130.

[35] Antonoplis, RA; Blanch, HW; Freitas, RP; Sciamanna, AF; Wilke, CR. *Biotechnol. Bioeng*. 25, (1983), 2757-2773.

[36] Erckel, R; Franz, R; Woernle, R; Riehm, T. (1985). Process for hydrolyzing cellulose-containing material with gaseous hydrogen fluoride. *US Patent*, 4, 556, 431.

[37] Bhat, MK; Bhat, S. *Biotechnol. Adv*. 15, (1997), 583-620.

[38] Onda, A; Ochi, T; Yanagisawa, K. *Green Chem*. 10, (2008), 1033-1037.

[39] Lima, S; Dias, AS; Lin, Z; Brandão, P; Ferreira, P; Pillinger, M; Rocha, J; Calvino-Casilda, V; Valente, AA. *Appl. Catal. A*, 339, (2008), 21-27.

[40] Moliner, M; Román-Leshkov, Y; Davis, ME (2010). Tin-containing zeolites are highly active catalysts for the isomerization of glucose in water. *Proceedings of the National Academy of Sciences of the United States of America*, Vol. 107, 6164-6168.

[41] Román-Leshkov, Y; Chheda, JN; Dumesic, JA. *Science*, 312, (2006), 1933-1937.

[42] Román-Leshkov, Y; Barrett, CJ; Liu, ZY; Dumesic, JA. *Nature*, 447, (2007), 982-985.

[43] Williams, CL; Chang, CC; Do, P; Nikbin, N; Caratzoulas, S; Vlachos, DG; Lobo, RF; Fan, W; Dauenhauer, PJ. *ACS Catal*. 2, (2012), 935-939.

In: Xylenes
Editor: Michael Olawale Daramola

ISBN: 978-1-62808-342-2

Chapter 4

INFLUENCE OF PLATINUM ON THE PERFORMANCE OF BETA ZEOLITE-BASED CATALYSTS FOR XYLENES PRODUCTION

Maria do Carmo Rangel[1,*], Saulo de Tarso Figueiredo Grecco[1], Diego Rodrigues de Carvalho[1], Ernesto Antonio Urquieta-González[2], Javier Mario Grau[3], Patricio Reyes[4] and Marcelo Oportus[4]

[1]GECCAT Grupo de Estudos em Cinética e Catálise, Instituto de Química, Universidade Federal da Bahia, Campus Universitário de Ondina, Salvador, Ba, Brazil
[2]Departamento de Engenharia Química, Universidade Federal de São Carlos, São Carlos, São Paulo, Brazil
[3]Instituto de Investigaciones en Catálisis y Petroquímica (INCAPE), (FIQ, UNL-CONICET), Santa Fe, Argentina
[4]Facultad de Ciências Químicas, Universidad de Concepción, Casilla 3-C Concepción, Chile

ABSTRACT

Xylenes are important raw materials with considerable application for chemicals and petrochemicals industry such as the production of synthetic fibers, plastics, resins and other products. The major sources of these compounds, naphtha reforming and gasoline pyrolysis, generate an appreciable amount of toluene and trimethylbenzene. The production of xylenes can be optimized by transalkylation of these compounds or by toluene disproportionation. The commercial process uses large pore zeolites which, however, deactivate with time mainly due to coke formation. In order to obtain new catalysts for these reactions, catalysts based on beta zeolite containing 0.5 and 1.0% platinum were synthesized and evaluated in toluene disproportionation. Zeolite-based materials were prepared by a two-step procedure and the metal was incorporated in the catalysts by wetness impregnation or by introducing platinum in the zeolite synthesis gel.

* E-mail: mcarmov@ufba.br.

The catalysts were characterized by differential thermal analysis, thermogravimetry, Fourier transformed infrared spectroscopy, X-ray diffraction, and adsorption/desorption of nitrogen and transmission electron microscopy. The activities of metallic and acidic sites were evaluated by cyclohexane dehydrogenation, a model reaction for both kinds of sites. All catalysts were more active than the commercial mordenite in toluene disproportionation, carried out at 470 °C and 1 atm. Platinum decreased the activity but increased the xylene selectivity, especially p-xylenes. The most promising catalyst was prepared by incorporating platinum (0.5%) by impregnation, which produces the highest amount of xylenes, mainly p-xylenes and m-xylenes, high value chemicals.

Keywords: Catalysis, para-xylenes, meta-xylenes,zeolite catalyts

INTRODUCTION

Aromatics play an important role in chemical and petrochemical industries, being used as raw materials for the manufacture of several high value chemicals. Benzene, toluene and xylenes (BTX) are among the most important ones, being extensively used to produce the majority of the intermediates used in the production of synthetic fibers, plastics and resins [1]. Among them, xylenes are by far the most valuable ones due to their wide range of applications in various fields, such as solvents and components in several final products. The individual isomers are even more valuable because of their applications. p-Xylene, for instance, is the precursor for the manufacture of terephthalic acid and dimethyl terephthalate both starting materials for the production of polyethylene terephthalate (PET), used in the manufacture of plastic bottles and polyester clothing. On the other hand, o-xylene is used for manufacturing phthalic anhydride and plasticizers, which are responsible for the flexibility of polyvinyl chloride (PVC). In addition, m-xylene can be oxidized to isophthalic acid to produce polyurethane and Nomex aramid polymer, [2-4]. Currently, p-xylene is responsible for about 83% of world demand for xylenes, which is higher than 42 million t/y [5].

Xylenes are commercially produced by naphtha reforming and as a by-product of naphtha cracking for ethylene production (gasoline pyrolysis), in which large amounts of C7 and C9 aromatics are also produced [1, 6, 7]. By upgrading these low value streams through aromatic transalkylation or disproportionation reactions, xylene production can be optimized, with the advantage of developing environmental friendly processes at low operating costs [7].

The toluene disproportionation occurs in equilibrium with isomerization and transalkylation of aromatics such as xylenes, trimethylbenzene and others [8]. However, the control of kinetic parameters, the reactivity of aromatics, the kind of catalytic sites and of the morphology and texture of the catalyst can affect the equilibrium shift towards the desired product. Therefore, these factors determine the occurrence of disproportionation or transalkylation reactions [9-11]. The disproportionation of aromatics leads to products of higher commercial value, mainly p-xylene. Toluene has the lowest market demand while benzene and xylenes present an annual growth rate of about 10% [12]. The process thus represents a promising alternative for the production of xylenes and benzene and may be performed in the presence of liquid or solid acid catalysts [1].

The first commercial process of alkylbenzenes disproportionation had begun in 1953, using a catalyst based on hydrofluoric acid and boron fluoride ($HF-BF_3$) [13]. In 1964, the process was improved by replacing the catalyst by a Friedel-Crafts type one, based on aluminum chloride [1]. However, due to environmental and technological problems related to the use of liquid acid catalysts, solid acids have become promising candidates to replace liquid systems in various industrial processes [14, 15]. Currently, most commercial processes for toluene disproportionation use catalysts based on ZSM-5, which deactivate with time mainly due to coke deposition. This process occurs due to pore blockage, since the first coke molecules formed in the large channels are retained because of their low volatility. Therefore, one coke molecule is enough to inhibit the diffusion of the reactant to the active sites of the channels [16, 17].

Due to the susceptibility of ZSM-5 catalyst to deactivate by coke deposition, new zeolites with larger pores (mordenite, Y, beta, SSZ-33 and TNU-9) have been extensively studied [18-20] as well as silica [21] and others. For most cases, the concentrations of xylene isomers produced were close to the thermodynamic equilibrium ratios. Since the amount of p-xylene obtained directly from disproportionation can not satisfy the market requirements, there is an urgent need to maximize the production of this isomer.

Several studies have shown [22-24] that the selectivity of p-xylene isomer was increased by optimizing the crystal diameter, Si/Al ratio, activation and pelletization methods for the zeolites. The most common method for improving p-xylene selectivity is to modify the catalysts with magnesium, phosphorous, boron, calcium, lanthanum and/or cerium oxides. These dopants act by adjusting the pore radii and/or the protecting the external surface sites against coke deposition. In general, these changes increase the p-xylene selectivity but decrease the activity resulting in low p-xylene yield.

Therefore, other catalysts have been studied for this reaction, such as micro-mesoporous composites based on zeolites [24-26] and metal-containing zeolites [27-31]. The micro-mesoporous composites have attracted much attention for toluene disproportionation to produce benzene and xylenes, since the reaction involves a bulky intermediate which requires a large pore size to be accommodated [26]. On other hand, zeolite-supported metals have been successfully used to improve toluene disproportionation. The highly dispersed metal on zeolite is known to have electron donor/acceptor ability, which stabilizes the conversion values and suppresses coke deposition. Bhavani et al. [29], for instance, have studied the toluene disproportionation and transalkylation of aromatic C9+ in the range between 370 and 550°C, in the presence of hydrogen using copper and/or palladium catalysts. They have found that the best composition for the catalyst is 8.7 wt% Cu and 0.69 wt% Pd supported on mordenite, which led to a conversion of 53% [29]. In another works [30, 31], we have studied the influence of nickel, palladium and platinum in transalkylation of trimethylbenzene with benzene and have found that the nickel-based catalysts produced mainly ethylbenzene while the platinum-based ones produced large amount of xylenes, being the best metal to improve the catalyst.

In the present work, we studied platinum-supported beta zeolite containing 0.5 and 1.0% platinum for toluene disproportionation, in order to obtain catalysts more selective to p-xylenes than the commercial one.

EXPERIMENTAL

Synthesis of Beta/MCM-41 Composite

The beta/MCM-41 composite was prepared by mesostructuration/crystallization of beta zeolite seeds, using a method based on the methodology previously reported but with some changes [25]. The beta zeolite seeds gel (1.47 Na_2O: Al_2O_3: 30 SiO_2: 8.23 TEAOH: 368 H_2O) was obtained by mixing silica, sodium aluminate, sodium hydroxide, tetraethylammonium hydroxide (TEAOH) and water. The gel produced was submitted to a hydrothermal treatment at 140°C, for 192 h. The seeds gel was mesostructured with cetyltrimethylammonium bromide (CTAB) by adjusting the pH between 9.0 and 9.5 and then was undergone to a hydrothermal treatment at 140°C, for 48 h. After this step, the sample was centrifuged, washed with water and dried at 60°C, for 24 h. The solid underwent to ion exchange with 300 mL of an aqueous solution of ammonium chloride (0.1 $mol.L^{-1}$) at 65°C, for 3 h. After this, the suspension was centrifuged, the solid was washed with water and dried at 60°C, for 24 h. The sample was then calcined at 550°C for 6 h for obtaining the zeolite in the protonic form (B sample).

Synthesis of Catalyst

The catalysts were obtained by dispersing the supports previously prepared in a hexachloroplatinic acid solution and keeping the suspension at rest, for 24 h. In order to obtain catalysts with 1.0% platinum, approximately 2 mL of the hexachloroplatinic acid solution (0.025 $mol.L^{-1}$) per gram of the support were used. Then, the solid was dried at 120°C, for 12 h and calcined at 350°C, for 3 h producing a platinum supported on beta/MCM-41 composite (1%PtB sample). For the material containing 0.5% platinum a different concentration (0.0125 $mol.L^{-1}$) of the hexachloroplatinic acid solution was used keeping the same conditions described, in order to obtain the 0.5%PtB sample.

Incoporation of Platinum in the Zeolite Synthesis Gel

The catalyst with 1% platinum was also prepared by adding the platinum precursor in the zeolite synthesis gel. In this case, the solid was prepared by adding 0.16 g of nitrate tetraminplatinum (II) dissolved in water during the preparation of zeolite seeds, just before of the addition of cetyltrimethylammonium bromide (CTAB). The same methodology previously described was then followed for getting the 1%PtBG sample.

Samples Characterization

The support was characterized by thermogravimetry, differential thermal analysis, Fourier transform infrared spectroscopy, X-ray diffraction and specific surface area and porosity measurements. The catalysts were characterized by specific surface area and porosity measurements, X-ray diffraction and transmission electron microscopy. The activities of metallic and acidic sites were evaluated by cyclohexane dehydrogenation, a model reaction for both kinds of sites.

The decomposition and the removal of the templates from the solids were monitored by thermogravimetry (TG) and differential thermal analysis (DTA). These experiments were carried out in a Mettler Toledo model TGA/SDTA 851e equipment, using 5 mg of the sample which was placed in a specimen holder of alumina and heated (10°C.min^{-1}) from room temperature to 1000°C under air flow (50 mL.min^{-1}).

The removal of the template and the production of the zeolite structure were also followed by Fourier transform infrared spectroscopy (FTIR). The experiments were performed in a Perkin Elmer model Spectrum One spectrophotometer. For each spectrum 32 scans were obtained in the range between 400 and 4000 cm^{-1} with a resolution of 4.0 cm^{-1}. The sample (about 50 mg) was mixed with 200 mg of potassium bromide and placed in a metal sample holder, with approximately 10 mm in diameter and 2.3 mm thick. All experiments were carried out at room temperature.

The formation of hexagonal mesoporous structure of MCM-41 type and of the microporous structure of beta zeolite in the samples was investigated through X-ray diffraction (XRD), performed in a Shimadzu model XRD 6000 equipment, using CuKα radiation generated at 40 kV and 30 mA and a monochromator. The analysis was performed in the 2θ scanning range from 1 to 5° with a goniometer speed of 0.5°.min^{-1}, for the mesoporous phase identification and from 5 to 45° with a goniometer speed of 2.0°.min^{-1}, for the microporous phase identification.

The textural properties of the catalysts were determined by nitrogen adsorption experiments performed at 77 K in a Micromeritics model ASAP 2020 apparatus. Before analysis, approximately 0.20 g of the sample was heated (10°C.min^{-1}) up to 250°C for 4 h under vacuum and then submitted to a pressure increase up to 10 μmHg, for the removal of water adsorbed on the sample and in the zeolite pore system. During analysis, the samples were submitted to pulses of nitrogen until a maximum increase of pressure of 925 mmHg.

The transmission electron microscopy and electron diffraction analysis were performed in a Jeol model JEM 1200 EXII equipment. The samples were dispersed in ethanol and one drop of the suspension was placed on carbon-coated copper grids (150 mesh). In the electron diffraction experiments, an accelerating voltage and the focal length 120 kV and 60 cm respectively were used.

The activities of the metallic and acidic sites were measured by the model reaction of cyclohexane dehydrogenation. The experiments were performed in a glass reactor at 300°C and 1 bar. Before reaction, the catalyst (0.1 g) was reduced under hydrogen flow (80 cm^{3}. min^{-1}) at 500°C, for 1 h. The products were analyzed in a Varian model 3400 CX chromatograph equipped with a FID and a packed column of FFAP on Chromosorb.

Catalysts Evaluation

The catalysts were evaluated in toluene disproportionation. The experiments were carried out in a tubular microreactor operating at 470 °C and 1 atm, using H_2/hydrocarbon (molar)= 4 and WHSV= 1.0 h^{-1}. Prior to the tests, the catalyst (0.5 g) was heated at 500 °C, under hydrogen flow, for 2 h. The gaseous effluent was analyzed by on line gas chromatography, using a Thermofinnigan model Trace GC instrument with FFAP capillary column and ionization detector. A commercial mordenite was also evaluated to be used as a reference. After each experiment, the reactor with the catalyst was cooled under nitrogen up to room temperature. The spent catalysts were analyzed by chemical analysis (carbon amount) in a CS-200 LECO model equipment using a ceramic crucible containing 0.20 g of sample, 1.25 g of tungsten compound (Lecocel) and 1.25 g of accelerator to help the combustion.

RESULTS AND DISCUSSION

The thermogravimetry curve of the beta/MCM-41 composite is shown in Figure 1a. In agreement with Kleitz et al. [32], this curve displayed four main stages of weight loss. The first one, at temperatures below 100°C, is related to the desorption of water while the second and third steps, between 150 and 500°C, can be associated to the decomposition and combustion of the templates in zeolite (TEAOH) and in MCM-41 (CTAB) structures. The weight loss at higher temperatures (near 600°C) is attributed to dehydroxylation of silanol groups and to combustion of residual coke.

The curve of differential thermal analysis (Figure 1b) shows that the decomposition of the templates involves three steps, in accordance with previous work [32]. It can be observed an endothermic peak near 250°C, related to the elimination of trimethylamine by Hofmann degradation, which leads to hydrocarbons formation. In the range from 250 to 350°C, they go into decomposition to produce lower molecular weight compounds. At this stage, it can be observed two exothermic events, each one associated to the decomposition of the hydrocarbon chain from each template. Finally, most of the oxidation processes occurs between 350 to 450°C, transforming the low molecular weight organic compounds into carbon dioxide, water and carbonaceous residues (exothermic process). The decomposition and the removal of the templates from the samples during calcination were also followed through FTIR. The spectra of the samples before calcination (Figure 2a) show bands related to symmetrical and asymmetrical stretching of C-H bonds near 2923 and 2853 cm^{-1}, respectively. These bonds are characteristic of CH_2 and CH_3 groups from the templates. After calcination, those bands disappeared, indicating that the calcination process was efficient for the template removal [33]. It can also be observed the band of angular deformation of CH bonds at 1454 cm^{-1} (δCH_3) and 1394 cm^{-1} (δCH_2) [33]. After calcination, these bands disappeared, indicating that the calcination process was effective for removing these compounds.

Figure 2b shows the spectra enlargement in the wavenumber range from 1400 to 400 cm^{-1} in order to make the visualization of the characteristic bands of amorphous and crystalline materials easier. It can be observed a band nearby 460 cm^{-1}, assigned to the internal vibrations of the aluminosilicate tetrahedra, which is insensitive to structural variations and thus is characteristic of both zeolitic and amorphous materials [34]. The spectra of the samples also exhibit two absorption band between 500 and 600 cm^{-1}, indicating the presence of a ring-like structure with five members of siloxan bond, T-O-T (T= Si, Al), typical of beta zeolite seeds [34]. The high crystallinity of the solids after calcination was detected by X-ray diffraction, as shown in Figure 3a, in which the Bragg reflections of beta zeolite in the high angle region can be observed [35]. No significant change was noted due to platinum incorporation in solids, indicating that the structure of beta zeolite was kept after impregnation. The X-ray diffraction profiles performed in the low angle region, from 1 to 10° 2θ (Figure 3b), displayed only the main diffraction peak (100) of the hexagonal mesoporous structure [36], which appeared broad and with very low intensity, in accordance with previous work [25]. The diffraction peak around 8 and 9° (2θ), shown in Figure 3b, is assigned to Bragg reflections (101) of beta zeolite. Only the pure support presented the other Bragg reflection (110) of the mesoporous structure. These results suggest that the beta/MCM-41 composite is made of two phases: (i) a crystalline one corresponding to the high stable beta zeolite and (ii) a poorly ordered mesophase of MCM-41 type [25]. For all cases, the zeolite structures were not significantly affected by impregnation, although a slight disordering of the mesoporous structure could be noted due to platinum incorporation.

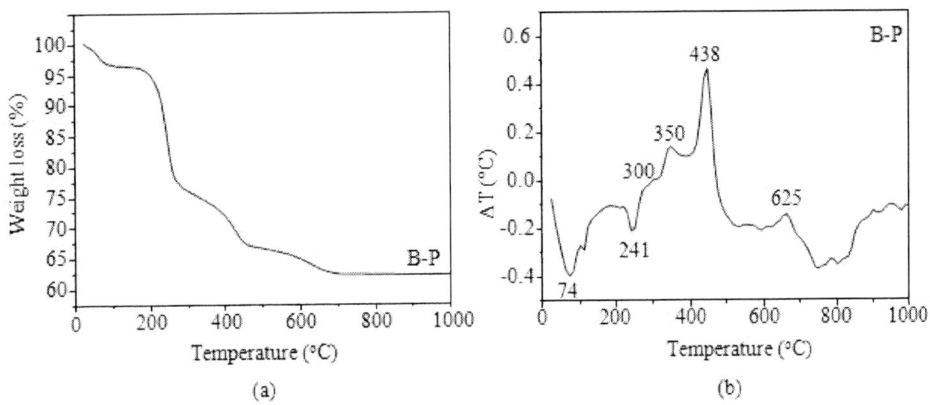

Figure 1. Curves of (a) thermogravimetry and of (b) differential thermal analysis for the precursor of beta/MCM-41 composite (B-P sample).

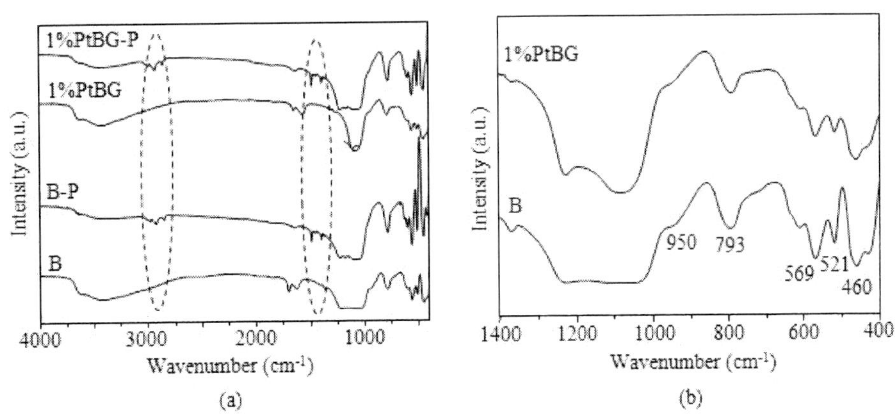

Figure 2. FTIR spectra of as synthesized (P) and calcined samples based on beta zeolite and (b) FTIR spectra of samples based on beta zeolite (B) enlargement in the wavenumber range from 1400 to 400 cm^{-1}. The letter G indicates the catalyst in which platinum was incorporated into the gel.

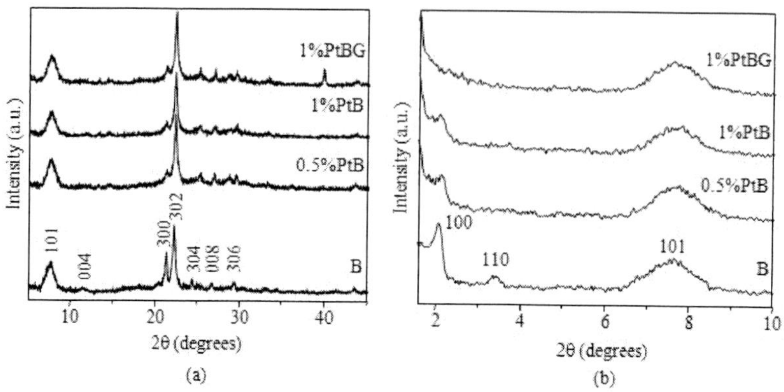

Figure 3. X-ray diffractograms performed (a) in the high angle region and (b) in the low angle region for calcined samples based on beta zeolite (B). The letter G indicates the catalyst in which platinum was incorporated into the gel.

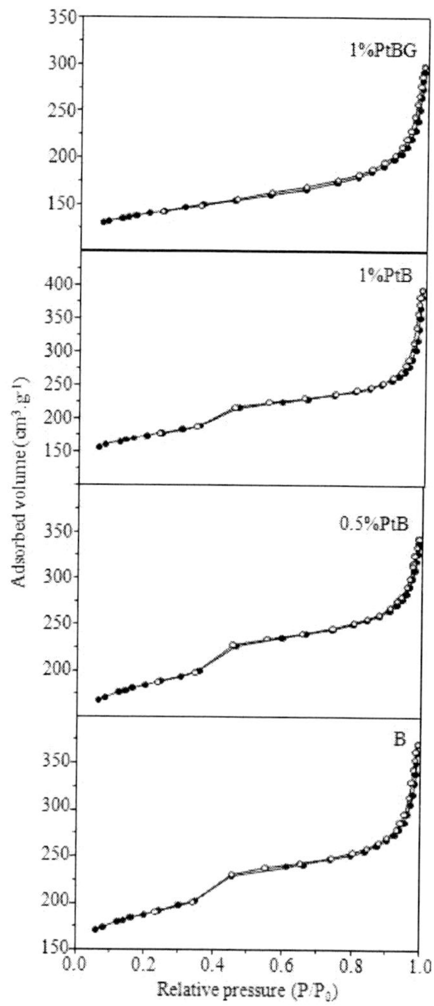

Figure 4. Nitrogen adsorption (-•-) and desorption (-o-) isotherms for calcined samples based on beta zeolite (B). The letter G indicates the catalyst in which platinum was incorporated into the gel.

As shown in Figure 4, all solids show isotherms, which are intermediates between I type, typical of microporous materials and IV type, typical of mesoporous materials [37]. As mentioned before, the presence of microporous is related to the structure of highly crystalline beta zeolites. Otherwise, the presence of mesoporous can be related to the formation of a poorly ordered mesophase during the mesostructuration step in the presence of CTAB and from mesopores among the particles, which are able to adsorb nitrogen at high relative pressures. This behavior shows that the synthesis of beta zeolites in two steps, used in the present work, results in a solid formed by small zeolite crystals between amorphous silica particles. In the first step, the synthesis conditions allow the formation of beta zeolite seeds whose sizes make their accommodation unfeasible around the CTAB micelles during mesostructuration; therefore, a less ordered mesoporous structure is generated, which become even lesser ordered as the crystals grow during the crystallization step. These results were confirmed by transmission electron microscopy that showed: (i) the presence of particles with irregular shape and heterogeneous characteristics, indicating that the solids are made of more than one phase (Figure 5a) and (ii) the dominion of mesophase (Figure 5b). The electron diffraction image corresponds to the crystalline dominion of beta zeolite (Figure 6).

The textural properties of the catalysts are shown in Table 1. As expected, they showed high specific surface areas and presented a narrow pore size distribution (Figure 7) with an average diameter of around 3.1 nm as shown by its pore size distribution determined by the BJH method from the adsorption branch of the nitrogen isotherms [38].

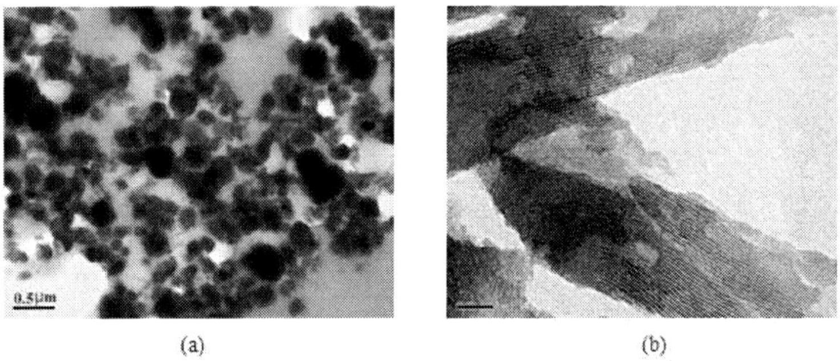

Figure 5. TEM images of particles of beta/MCM-41 composite showing (a) the different phases of the solids and (b) the domain of mesophase.

Figure 6. Electron diffractogram of the crystalline phase of beta zeolite for the beta/MCM-41 composite.

Table 1. Specific surface area obtained by BET method (Sg), micropore area obtained by *t*-plot method (Sg_m), mesopore area with diameter between 2 and 50 nm obtained by BJH method (Sg_{mp}), micropore volume (V_m), mesopore volume (V_{mp}) and mesopore diameter (dp) for the samples based on beta zeolite (B). The letter G indicates the catalyst in which platinum was incorporated into the gel

Samples	Sg ($m^2.g^{-1}$)	Sg_m ($m^2.g^{-1}$)	Sg_{mp} ($m^2.g^{-1}$)	V_m ($cm^3.g^{-1}$)	V_{mp} ($cm^3.g^{-1}$)	dp (nm)
B	644	392	202	0.1811	0.2649	3.1
0.5%PtB	631	384	203	0.1778	0.2516	3.1
1%PtB	597	350	207	0.1619	0.2879	3.1
1%PtBG	476	310	98	0.1446	0.1921	3.2

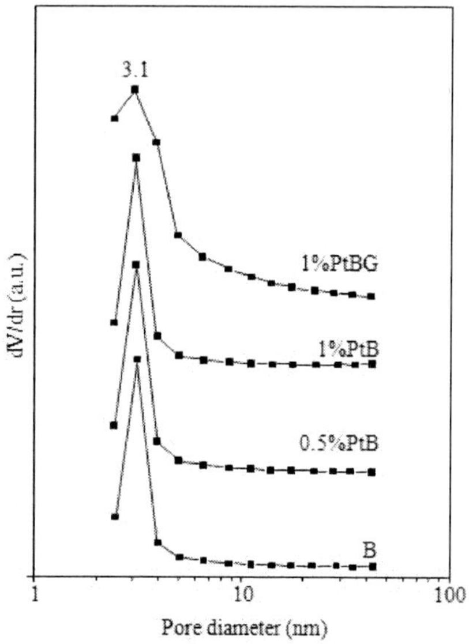

Figure 7. Pore size distributions for the samples based on beta zeolite (B). The letter G indicates the catalyst in which platinum was incorporated into the gel.

Platinum incorporation did not significantly affect the textural properties of the solids, a fact that can be attributed to the low amounts of the metal. On the other hand, these properties are strongly affected by the introduction of metal into the beta zeolite seeds gel. This finding suggests that the metal is located inside the micro and mesopores of this solid.

The catalysts were active in cyclohexane dehydrogenation, as shown in Figure 8. As we can see, they showed different activities in the beginning of reaction, being the catalyst with the lowest platinum content the most active one followed by the 1%PtB and 1%PtBG samples. These results are closely related to benzene selectivities (Figure 9a), indicating that platinum dispersion follows this order. Therefore, the amount of platinum as well as the method of its incorporation in the solid affect the metal dispersion. On the other hand, the catalysts showed similar selectivities to methylcyclopentane (Figure 9b), indicating that they have similar acidities and then the platinum amount and the method of its incorporation in the solid have no effect on the acidity.

All samples were active in toluene disproportionation (Figure 10 and Table 2). It can be noted that the platinum-richest catalyst prepared by impregnation was the most active one, followed by beta zeolite. On the other hand, the platinum-poorest sample was the least active. However, the catalysts deactivated during the first two hours of reaction, which can be related to the coke produced on the catalysts (Table 2). After 7 h of reaction the conversion achieved stable values, which were higher than commercial mordenite, in spite of the coke. This finding can be related to the tridimensional micropore structure and mesoporosity of the beta zeolite-based catalysts that allow the access of the reactant to the active site inside the pores even in the presence of coke, partially blocking the pores. The catalysts are selective to different products, as shown in Table 2. As we can see, the samples based on beta zeolite were more active than the commercial mordenite, after 10 h on stream. The addition of low amounts of platinum (0.5%) to the catalyst did not affect the activity but higher amounts caused a decrease, the incorporation of the metal in the gel led to the largest decrease. The metal strongly affected the selectivity to reaction products. The platinum-containing catalysts showed high selectivities to xylenes, followed by benzene and ethylbenzene, indicating that toluene disproportionation occurred to a greater extent than transalkylation. The highest selectivity to xylenes was shown by the catalyst with 0.5%Pt. The distribution of xylenes isomers was also affected both by platinum and its incorporation method in solids, the highest amount of p-xylenes was produced on the catalyst with 1% Pt and prepared by impregnation. The highest xylenes and p-xylene yields were shown by the catalyst with 0.5%Pt.

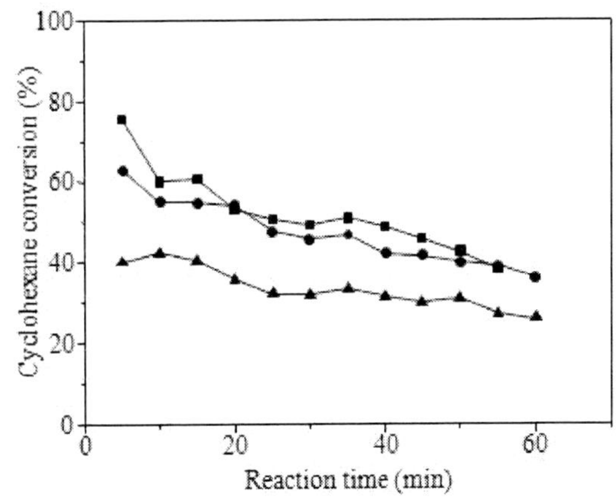

Figure 8. Cyclohexane conversion as a function of time on stream at 300°C over the catalysts based on beta zeolite (B) during cyclohexane dehydrogenation. -■- 0.5%PtB, -●- 1%PtB and -▲-1%PtBG.

Table 2. Catalytic activity (a), conversion (X_T), selectivity to benzene (S_{BZ}), ethylbenzene (S_{EB}), p-xylene ($S_{P\text{-}X}$) m-xylene ($S_{m\text{-}X}$), o-xylene ($S_{O\text{-}X}$), total xylenes (S_{XT}) during disproportionation of toluene and coke (C) on the spent catalysts based on beta zeolite (B), after 10 h on stream. CM= commercial mordenite

Samples	a x 10^5 ($mol.g^{-1}s^{-1}$)	X_T (%)	S_{BZ} (%)	S_{EB} (%)	$S_{p\text{-}X}$ (%)	$S_{m\text{-}X}$ (%)	$S_{o\text{-}X}$ (%)	S_{XT} (%)	C (%)	Y_X (%)	$Y_{p\text{-}X}$ (%)	$Y_{o\text{-}X}$ (%)
CM	3.3	11	9.0	5.4	21	8.5	2.9	32	1.7	3.5	2.3	0.32
B	8.4	28	23	1.3	14	0.0	4.3	18	14	5.0	3.9	0.36
0.5%PtB	8.7	29	7.7	0.0	30	0.0	33	62	14	18	8.7	9.6
1%PtB	4.8	16	28	3.7	43	0.0	14	57	14	9.1	7.0	2.2
1%PtBG	3.6	12	48	5.6	32	0.0	10	42	13	5.0	3.8	1.2

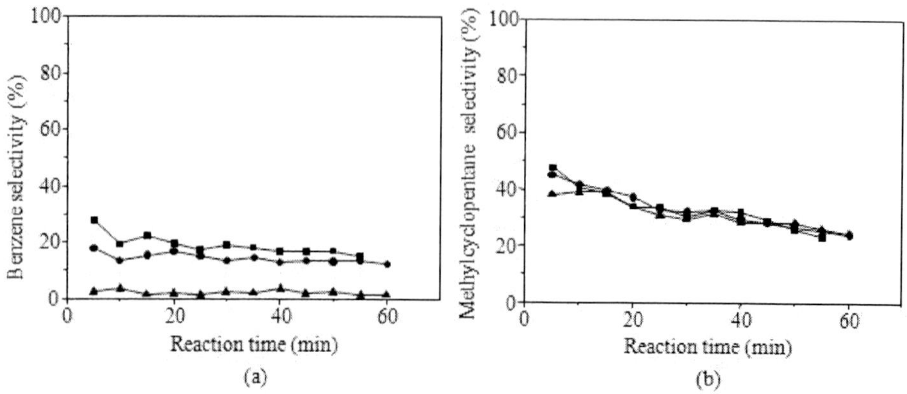

Figure 9. Selectivities to (a) methylcyclopentane and to (b) benzene for the catalysts based on beta zeolite (B) in cyclohexane dehydrogenation as a function of time on stream at 300°C. -■- 0.5%PtB, -●- 1%PtB and -▲-1%PtBG.

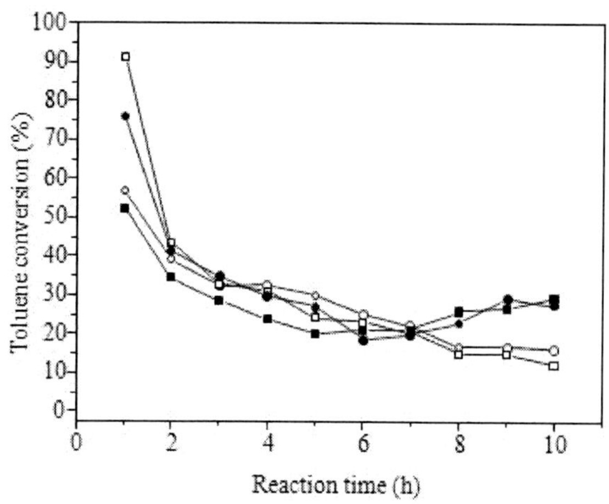

Figure 10. Toluene conversion over the catalysts based on beta zeolite (B) during toluene disproportionation as a function of time on stream at 470°C. -●- B, -■- 0.5%PtB and -□- 1%PtB, -o- 1%PtBG.

Conclusion

The beta/MCM-41 composite prepared by two steps is made of a microporous phase corresponding to beta zeolite with high crystallinity and a poorly ordered mesoporous phase related to MCM-41, which provides high specific surface area to the solid. The addition of platinum (0.5 and 1%) does not affect the textural properties of the composite, except when introduced directly into the zeolite seeds gel. These catalysts are more active than the commercial mordenite in toluene disproportionation. The addition of platinum strongly affects the catalytic activity and selectivity of beta/MCM-41 composite, decreasing the activity and increasing the selectivity to xylenes, especially to p-xylene. The distribution of xylenes isomers was also affected by platinum as well as by its incorporation method in solids; in general, the catalysts were more selective to p-xylene, followed by o-xylene, the most valuable isomers and did not produce m-xylene. The most suitable catalyst to produce xylenes, especially p-xylenes, was prepared by incorporating platinum (0,5%) by impregnation.

REFERENCES

[1] Tsai, TC; Liu, SB; Wang, I. *Appl. Catal. A: General*, 181, (1999), 355-398.

[2] Olah, GA; Molnár, A. *Hydrocarbon Chemistry*, John Wiley & Sons, New Jersey, 2003.

[3] Sherman, JD. *Proc. Natl. Acad. Sci. USA*, 96, (1999), 3471-3478.

[4] Hocking, MB. *Handbook of Chemical Technology and Pollution Control*, third ed., Academic Press, New York, 2005.

[5] Balasamy, RJ; Odedairo, T; Al-Khattaf, S. *Appl. Catal. A: General*, 409-410, (2011), 223-233.

[6] Tsai, TC; Chen, WH; Liu, SB; Tsai, CH; Wang, I; Ku, CS. *Catal. Today*, 73, (2002), 39-47.

[7] Forni, L; Cremona, G; Missineo, F; Bellusi, G; Perego, C; Pazzuconi, G. *Appl. Catal. A: General*, 121, (1995), 261-272.

[8] Serra, JM; Guillon, E; Corma, A. *J. of Catal.* 227, (2004), 459-469.

[9] Tsai, TC. *Appl. Catal. A: General*, 301, (2006), 292-298.

[10] Tsai, TC; Chen, WH; Lai, CS; Liu, SB; Wang, I; Ku, CS. *Catal. Today*, 97, (2004), 297-302.

[11] Mavrodinova, V; Popova, M; Mihályi, RM; Pál-Borbély, G; Minchev, C. *J. of Molec. Catal. A: Chemical*, 220, (2004), 239-246.

[12] Harvan, RG. *Oil Gas J.* 30, (1998), 56-61.

[13] Lien, AP; Mccaulay, DA. *J. Am. Chem. Soc.* 75, (1953), 2407-2410.

[14] Degnan Jr., TF; Smith, CM; Venkat, CR. *Appl. Catal. A: General*, 221, (2001), 283-294.

[15] Sulikowski, B; Rachwalik, R. *Appl. Catal. A: General*, 256, (2003), 173-182.

[16] Rangel, MC; Valentini, A; Oliveira, AS; David, JM; Britto, JM; Domingues, SM; Reyes, P. *Quim. Nova*, 26, (2003), 305-308.

[17] Guisnet, M; Magnoux, P. *Appl. Catal. A: General*, 212, (2001), 83-96.

[18] Cavani, F; Corazzari, M; Bencini, E; Goffredi, G. *Appl. Catal. A: General*, 226, (2002), 31-40.

[19] Oliveira, AC; Essayem, N; Tuel, A; Clacens, J-M; Taarit, YB; Rangel, MC. *Stud. Surf. Sci. Catal.* 158, (2005), 1677-1684.

[20] Baduraig, A; Odedairo, T; Al-Khattaf, S. *Top. Catal*, 53, (2010), 1446-1456.

[21] Dias, JA; Rangel, MC; Dias, SCL; Caliman, E; Garcia, FAC. *Appl. Catal. A: General*, 328, (2007), 189-194.

[22] Uguina, MA; Sotelo, JL; Serrano, DP. *Appl. Catal.* 76, (1991), 183-198.

[23] Uguina, MA; Sotelo, JL; Serrano, DP; van Grieken, R. *Ind Eng Chem Res*, 31, (1992), 1875-1880.

[24] Mitra, B; Kunzru, D. *Catal. Lett.* 141, (2011), 1569-1579.

[25] Grecco, STF; Nobre, PSS; Urquieta-González, EA; Rangel, MC. in: A. Sayari, M. Jaroniec, (Eds.), Nanoporous Materials: Proceedings of the 5th International Symposium, World Scientific Publishing Co. Pte. Ltd. New Jersey, 2008, pp. 599-606.

[26] Ungureanu, A; Hoang, TV; Trong On, D; Dumitriu, E; Kaliaguine, S. *Appl. Catal. A: General*, 294, (2005), 92-105.

[27] Oliveira, AC; Rangel, MC; Fierro, JLG; Reyes, P; Oportus, M. *Quim. Nova*, 28, (2005), 37-41.

[28] Oliveira, AC; Reyes, P; Oportus, M; Rangel, MC. *Stud. Surf. Sci. Catal.* 154, (2004), 886-893.

[29] Bhavani, AG; Karthekayen, D; Rao, AS; Lingappanet, N. *Catal. Lett.* 103, (2005), 89-100.

[30] Ramos, MS; Grecco, ST; Gomes, LP; Oliveira, AC; Reyes, P; Oportus, M; Rangel, MC. *Stud. Surf. Sci. Catal.* 156, (2005), 809-814.

[31] Grecco, STF; Gomes, LP; Reyes, P; Oportus, M; Rangel, MC. *Stud. Surf. Sci. Catal.* 158, (2005), 1937-1944.

[32] Kleitz, F; Schimidt, W; Schüth, F. *Microporous Mesoporous Mater.* 65, (2003), 1-29.

[33] Selvaraj, M; Pandurangan, A; Seshadri, KS; Sinha, PK; Lal, KB. *Appl. Catal. A: General*, 242, (2003), 347-364.

[34] Flanigen, EM. in: J.A. Rabo, (Eds.), *Zeolite Chemistry and Catalysis*, American Chemical Society Washington, DC, 1976, 80-117.

[35] Treacy, MMJ; Higgins, JB. *Collection of Simulated XRD Powder Patterns for Zeolites*, fifth ed., Elsevier, Amsterdam, 2007.

[36] Beck, JS; Vartuli, JC; Roth, WJ; Leonowicz, ME; Kresge, CT; Schmitt, KD; Chu, CT-W; Olson, DH; Sheppard, EW; McCullen, SB; Higgins, JB; Schlenker, JL. *J. Am. Chem. Soc.* 114, (1992), 10834-10843.

[37] Schneider, P. *Appl. Catal. A*, 129, (1995), 157-165.

[38] Barrett, EP; Joyner, LG; Halenda, PP. *J. Am. Chem. Soc.* 73, (1951), 373-380.

In: Xylenes
Editor: Michael Olawale Daramola
ISBN: 978-1-62808-342-2

Chapter 5

NANOCOMPOSITE MFI-TYPE ZEOLITE MEMBRANE FOR SEPARATING XYLENE ISOMERS

M. O. Daramola[1,2], Z. Deng[3], A. J. Burger[4] and A. Giroir-Fendler[5]

[1]School of Chemical and Metallurgical Engineering, Faculty of Engineering & the Built Environment, University of the Witwatersrand, Johannesburg, South Africa

[2]Catalysis Engineering Section, Department of Chemical Engineering, Delft University of Technology, Delft, The Netherlands

[3]Chengdu Institute of Organic Chemistry, Chinese Academy of Science, Chengdu, P. R. China

[4]Department of Process Engineering, University of Stellenbosch, Stellenbosch, South Africa

[5]Engineering Research Group, Institute for Research on Catalysis & Environment (IRCELyon), 1 Albert Einstein, Villeurbanne Cedex, Lyon, France

ABSTRACT

The principal industrial sources of xylene isomer mixtures are high-severity catalytically reformed naphtha and pyrolysis distillates. The C_8 aromatic cut obtained from these sources contains a mixture of xylenes (50-60 wt.% meta-xylene and 20-25 wt.% o- and para-xylene) and ethylbenzene (EB) (15-30 wt.%) in the C_8 fractions obtained from naptha reforming and steam cracking. The use of distillation is discouraged for para-xylene separation and purification due to the close boiling points of xylene isomers, translating into high energy demands. For example, petrochemical and chemical industries accounted for 13.7 quintillion Joule (J) in 1998 with about 35% of the energy consumption used in the manufacture and separation of organic chemicals (mainly for heating/cooling). Moreover in 2004, the United States consumed nearly 105 quintillion Joule (J) on petrochemical and chemical industries' processes, corresponding to approximately one fourth of the world energy demand. Therefore, it becomes imperative to move to more energy-efficient and environmentally-friendly processes for para-xylene separation and purification involving the lowest number of heating/cooling steps. In this chapter application of membrane-based technology in the synthesis of xylenes, in particular the use of nanocomposite MFI-type (silicalite-1) zeolite membranes to separate xylene isomers into high purity para-xylene is discussed.

Keywords: Xylenes, MFI, nanocomposite membranes, zeolites, separation

INTRODUCTION

The principal industrial sources of xylene isomer mixtures are high-severity catalytically reformed naphtha and pyrolysis distillates. The C_8 aromatic cut obtained from these sources contains a mixture of xylenes (50-60 wt.% meta-xylene and 20-25 wt.% o- and para-xylene) and ethylbenzene (EB) (15-30 wt.%) in the C_8 fractions obtained from naptha reforming and steam cracking [1]. The surplus ortho-xylene and meta-xylene can be converted into more valuable para-xylene through catalytic isomerization with further purification by using convenient separation techniques. The use of distillation is discouraged for para-xylene separation and purification due to the close boiling points of xylene isomers (Table 1), translating into high energy demands. For example, petrochemical and chemical industries accounted for 13.7 quintillion Joule (J) in 1998 with about 35% of the energy consumption used in the manufacture and separation of organic chemicals (mainly for heating/cooling) [2] Moreover, in 2004, the United States consumed nearly 105 quintillion Joule (J) on petrochemical and chemical industries' processes, corresponding to approximately one fourth of the world energy demand [3]. Therefore, it becomes imperative to move to more energy-efficient and environmentally-friendly processes for para-xylene separation and purification involving the lowest number of heating/cooling steps.

The existing commercial technologies for separation and production of high purity para-xylene from its isomers can be divided into three main groups: (1) fractional crystallization, (2) adsorption and (3) hybrid crystallization/adsorption [4]. For many years, first commercial technique for separating xylene isomers (para-xylene from mixed xylenes) employed low temperature fractional crystallization process and a number of crystallization processes was commercialized over the years (e.g., Chevron, Krupp, Amoco, ARCO [Lyondell] and Phillips). Example of a commercial process based on crystallization process is depicted in Figure 1. This technology exploits the difference in the freezing points of the isomers (see Table 1). Since para-xylene possesses the highest freezing point when compared with other two isomers, a pure solid phase of para-xylene crystallizes first upon cooling to ~286 K. Then the crystallized para-xylene crystals are separated from the mother liquor by either filtration or centrifugation. However, achieving good separation depends on the para-xylene crystal size distribution. The para-xylene crystal size is thus affected by the degree of supersaturation upon crystallization, which is affected in turn by a number of parameters including temperature, agitation and the presence of crystal nuclei. Therefore to achieve high purity para-xylene, separation of the crystallized para-xylene is done in more than one or two consecutive steps of filtration and followed by centrifugation [5].

Adsorption technique is the second and the most recent method for separating and producing high purity para-xylene. In this process, adsorbents such as molecular sieves are used to produce high purity para-xylene by preferential adsorption of para-xylene from a mixed xylene stream. Separation is achieved by exploiting the differences in affinity of the adsorbent for para-xylene relative to the affinity of the adsorbent for other C_8 isomers. The adsorbed para-xylene is subsequently desorbed by using a desorbent liquid stream that extract para-xylene from the adsorbent. Typical recovery per pass in a single step with adsorption technique is ~95%. For adsorption technique, several adsorbent/desorbent combinations have been proposed in the literature to promote para-xylene recovery from different mixtures [7,8]. For example, Ba-and K-exchanged zeolite molecular sieves have been used as adsorbents, and desorbent like toluene (or tetraline) has been used to recover para-xylene from the adsorbent [9]. However, the crystallization and adsorption techniques for para-xylene separation and purification are highly energy intensive (although to a lesser extent than in fractional distillation) and make use of batch processes. In an attempt to design less energy-intensive and more environmental benign processes for para-xylene synthesis, membrane-based technology has been proposed and studied as a promising alternative.

Table 1. Physical properties of xylene isomers

Xylene isomers	M_w ($g.mol^{-1}$)	Normal Boiling point (K)	Normal Freezing point (K)	ΔH_{vap} ($kJ.mol^{-1}$)
para-xylene	106.17	411.37	286.26	42.04
meta-xylene	106.17	412.12	225.13	42.04
ortho-xylene	106.17	417.41	247.82	43.41

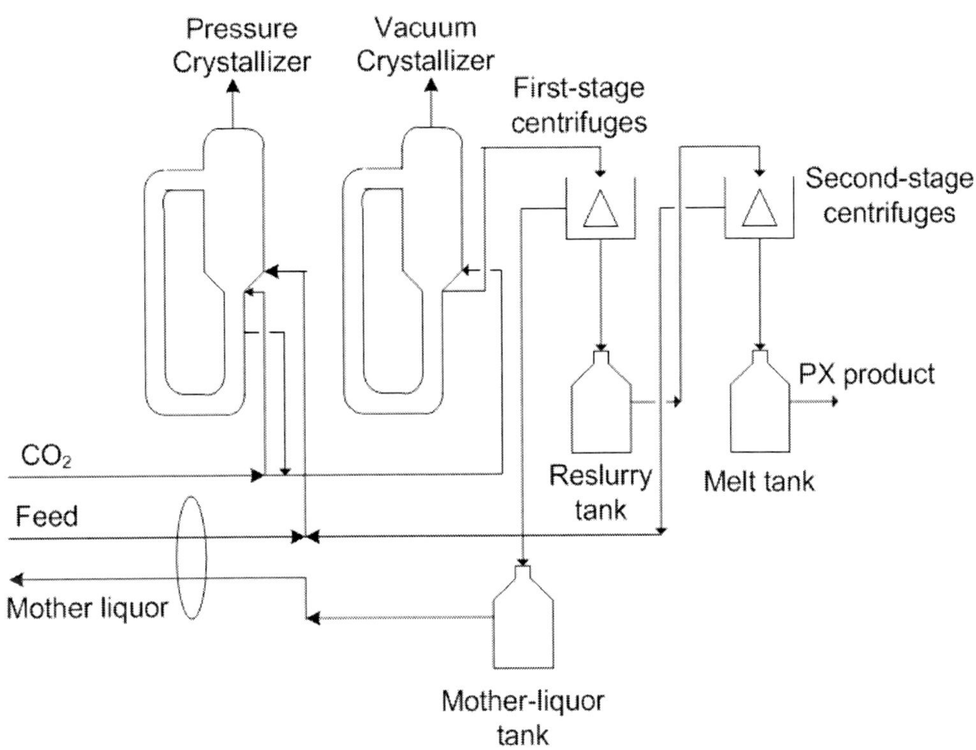

Figure 1. A typical commercial crystallization process for xylene synthesis (adapted from ref. [6]).

Membrane Technology for Separation Applications

Nowadays, membrane technologies are becoming more frequently used for the separation of wide range of mixtures in the petrochemical-related industries and can compete successfully with traditional schemes due to the following reasons [10-14]:

- The technology behind membrane-based separation is potentially an energy-saving one, because the separation process takes place without phase transition (for instance in gas separation but not in all cases)
- It is environmentally benign, since the membrane approach requires the use of relatively simple and non-harmful materials
- Minor but valuable components can be recovered from a main stream using membrane-based processes without substantial additional energy costs
- Membranes can offer a simple, easy-to-operate, low-maintenance process option when compared with existing conventional processes.
- Membrane-based processes have a low capital cost, compact size, modular configuration, and low specific power consumption, thereby reducing the production cost
- Contrary to conventional processes like adsorption and absorption processes, membrane-based processes do not require regeneration, a step leading to the use of two solid beds or a solvent regeneration unit.

A membrane is defined essentially as a barrier, which separates two phases and restricts transport of various chemicals in a selective manner. A typical membrane-based process is schematically described in Figure 2.

A membrane can be homogenous or heterogeneous, symmetric or asymmetric in structure, solid or liquid; it can either carry a positive or negative charge or it can be neutral or bipolar. General classification of membranes is depicted in Figure 3. Unlike most chemical engineering separation processes which are governed by phase equilibrium relations, membrane separation is based primarily upon the relative rates of mass transfer. Transport occurs by a solution-diffusion mechanism and membrane selectivity is based upon the relative permeation rates of the components through the membrane. Each component (gas/vapour) transporting through the membrane has a characteristic permeation rate that is a function of the ability to dissolve and diffuse. The two relationships upon which the equations are based are Fick's Law (diffusion) and Henry's Law (solubility). The transport through a membrane can be affected by convection or by diffusion of individual molecules, induced by an electric field or concentration, pressure or temperature gradient. Stages involved in the transport of a molecule through membranes are:

- adsorption of the molecule onto the interface of the high-pressure side of the membrane
- dissolution of the molecule into the membrane at the interface and diffusion of the molecule through
- the elution of the molecule from the membrane at the interface
- desorption of the molecule from the low pressure side of the membrane

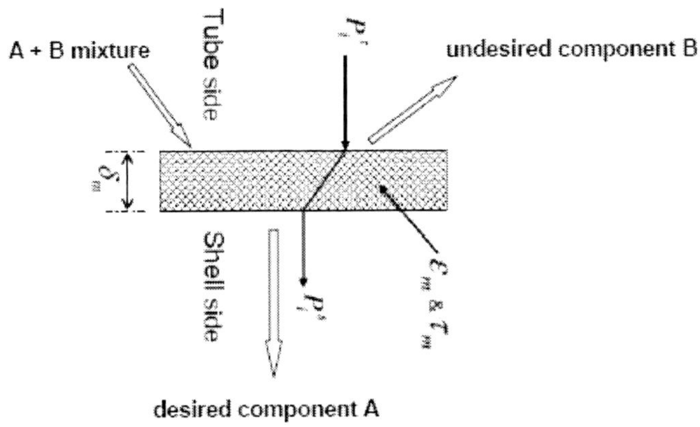

Figure 2. Schematic description of a membrane-based process. ε_m is the porosity of the membrane; τ_m, the membrane tortuosity [-]; δ_m, the membrane effective wall thickness [m]; J_i, the cross-membrane flux (mol.s^{-1}.m^{-2}). P_i^t and P_i^s is the partial pressures [Pa] of component i in the tube side and permeate side, respectively.

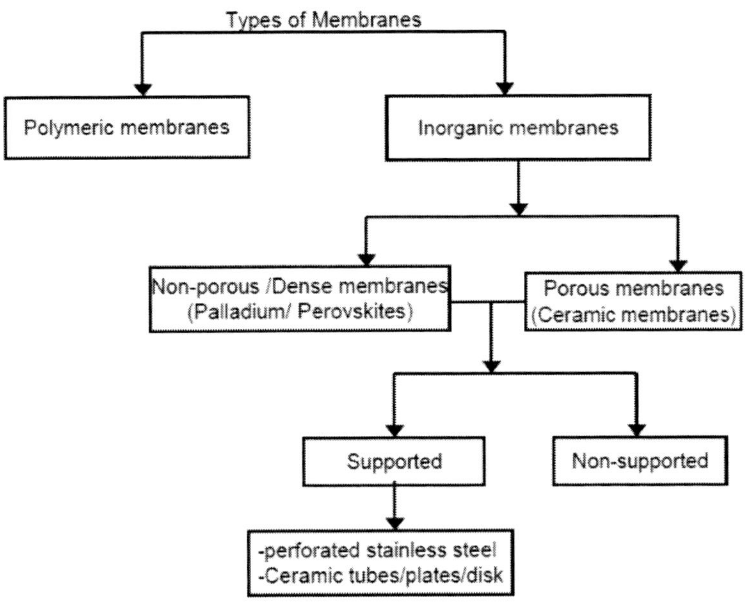

Figure 3. General classifications of membranes.

Polymeric Membranes

Application of polymeric membranes for hydrocarbon separation has been reported by several researchers [15-21]. Applications of glassy polymers for separation of olefins, paraffins aromatic, alicyclic and aliphatic hydrocarbons have been reported [15-17]. A few of research efforts have been expended on the application of rubbery polymers and polyimides for selective separation of hydrocarbon vapours from their mixtures as well as in pervaporation processes for the extraction of hydrocarbons from their aqueous [22,23].

Seperation of benzene/cyclohexane mixtures, toluene/iso-octane mixtures, aromatic/alicyclic/aliphatic hydrocarbon mixtures with DSDA-TrMPD/DEB (DEB-2,2'-diethynylbenzidine)copolyimide has been reported [22-25]. Furthermore, separation of xylene isomers with polyurethane, polyvinyl alcohol (PVA), polyamides [26-30], but none of the aforementioned membranes displayed reasonable selectivity towards para-xylene. Beside very low selectivity displayed by polymeric membranes, low resistance to chemicals, low thermal and mechanical stability rule out the application of polymeric membranes for gas separation, in particular xylene separation. As a result of the aforementioned shortcomings of polymeric membranes combined with the quest for energy efficient process for industrial separation of chemicals, research efforts have been focused on the development of inorganic/ceramic membranes for gas separation. A comparison between polymeric membranes and inorganic membranes is presented in Table 2.

MFI-Type Zeolite-Based Membranes

Zeolites are tridimensional crystalline aluminosilicates constituted by Si and Al tetrahedral linked through bridging oxygen atoms giving rise to the so-called *secondary building units* (SBUs), constituted by rings and prisms of various sizes (see Figure 4). These units combine to generate frameworks with a regular distribution of molecular-sized pores and cavities. The general formula of zeolite is $M_{x/n}[(AlO_2)_x\ (SiO_2)_y].zH_2O$, with M defining the compensating cation (usually from groups I or II) with valence *n*. The Si/Al ratio of the zeolite structure and amount of cations control the surface properties of zeolite (e.g., hydrophobicity and acidity), and determine their adsorbent, catalytic and ion-exchange properties.

Amongst all available natural and synthetic zeolite materials, MFI-type zeolite is the most studied zeolite material for xylene synthesis. The MFI-type zeolite structure consists of two main types: (i) ZSM-5 (Zeolite Socony Mobil-Five) and, (ii) silicalite-1. ZSM-5 was discovered by Mobil in 1972 [32,33] and it has a natural analog (mutinaite) [34]. Silicaite-1 was developed by Union Carbide a few years later after the discovery of ZSM-5 zeolite [35,36]. The two MFI-type zeolite structures differ in their Si/Al ratio, with Si/Al ratio = 10-10000 for ZSM-5 and Si/Al ratio >10000 for silicalite-1. Silicon content of the structure provides thermal stability and hydrophobicity characteristics of the zeolite, making silicate-1 more thermally stable and more hydrophobic than ZSM-5. The elementary mesh in MFI-type zeolites contains 96 TO_4 tetrahedic units (T = Si or Al) and it has the chemical formula shown in (1)[37].

$$[M_{x/n}(H_2O)_{16}][Al_x\,Si_{96-x}O_{192}] \text{ with } x < 27 \tag{1}$$

MFI-type zeolite (silicalite-1 or ZSM-5) possess two types of channel namely, near-circular straight (0.54x0.56 nm^2) and elliptical sinusoidal (0.51x0.55 nm^2) channels. The channels are defined by 10-membered rings (medium-pore zeolite) [38,39]. Straight and elliptical channels run parallel to crystallographic directions *a*(1 0 0) and *b*(0 1 0),respectively, generating perpendicular intersections with diameter of 0.89 nm. The key morphological features of MFI-type zeolite are schematically depicted in Figure 5. The pores provide three potential adsorption sites: (i) at the intersections, (ii) along the cross (zig-zag) channels and (iii) in the straight channels between the intersections. Considering the pore sizes of the channel network of MFI-type zeolite, the zeolite is expected to promote diffusion of para-xylene (kinetic diameter=0.58 nm) over ortho-xylene and meta-xylene isomers (kinetic diameter=0.68 nm each). These unique shape selectivity properties make MFI-type zeolite an excellent candidate zeolite for selective separation of para-xylene from xylene isomers or mixed xylenes.

Table 2. Comparison between polymeric and inorganic membranes

Inorganic membranes	Polymeric membranes
Do not swell	Do swell
Possibility of uniform, molecular sized pores allowing for molecular sieving	Do not have uniform molecular sized pores
Chemically resistant to solvents and low pH	Not chemically stable. Denatured at low pH
Thermally stable	Not thermally stable, denatured at high temperature
High cost of production	Lower cost of production
More brittle	Less brittle

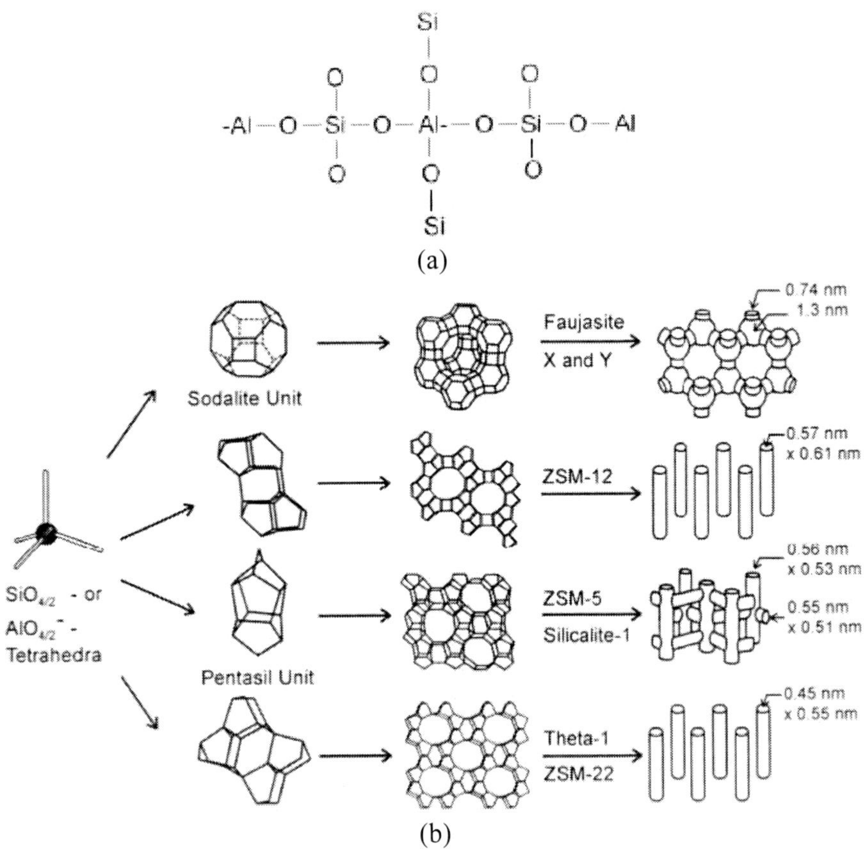

Figure 4. Structure of zeolite. (a) Basic zeolite structure (b) development of zeolite structures (from aluminosilicates to secondary building blocks to zeolite structures) showing zeolite X & Y; zeolite ZSM-5 or silicalite-1; zeolite ZSM-12 and zeolite Theta-1 or ZSM-22 (adapted from [31]).

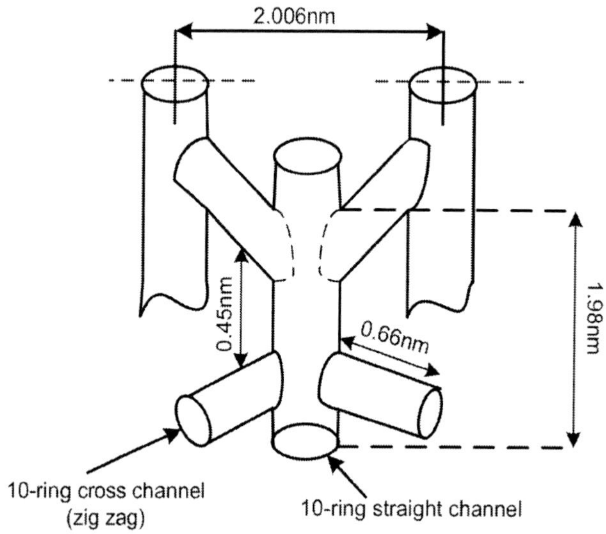

Figure 5. Morphological features of MFI-type zeolite showing near-circular straight (0.54x0.56 nm^2) and elliptical sinusoidal (0.51x0.55 nm^2) channels (adapted from ref.[5]).

Zeolites have been the major materials employed in the preparation of ceramic membranes for the separation of aromatics and several reviews on their synthesis and applications have been published [5,40-43]. Although the use of zeolite membranes for gas separation is still in an early technological stage (only one application of solvent dehydration by pervaporation has been commercialized to date [44], they are expected to encounter applications at large scale and compete with other existing technologies in the coming years. The future gas separation applications will rely on their selectivity, permeability and stability characteristics. Moreover, the development of such materials will depend on the availability of high quality, defect-free membranes, preferentially synthesized through simple protocols leading to cost-effective processes [45].

Preparation of Zeolite-Coated Membranes

Several strategies have been proposed and used to prepare zeolite-coated membranes [46-48]. Zeolite-coated membranes consist of a thin film of zeolite layer coated on a porous support (e.g. porous stainless steel, α-alumina), herewith referred to as thin-film zeolite (TFZ) membranes. Basically, all methods described in literature for fabricating TFZ membranes involve either the use of zeolite crystals previously synthesized or crystallization of zeolite layers. Figure 6 describes strategies involved in the preparation of zeolite membranes. Among them, dry-gel conversion and liquid-phase hydrothermal synthesis are commonly used for the synthesis of zeolite films onto a porous support in one or several batch cycles. Liquid-phase hydrothermal synthesis method consists of the deposition of a layer containing the Si and Al precursors as a dry amorphous aluminosilicate gel onto the support using sol-gel techniques, followed by zeolitization under the presence of vapours [49,50]. For example, TFZ of MFI-type zeolite membranes are often prepared by liquid-phase hydrothermal synthesis (otherwise known as direct in-situ hydrothermal synthesis) in an autoclave at 443-473 K under autogeneous pressure using an organic template as structure-directing agent (SDA), usually tetrapropylammoniunhydroxide (TPAOH) and/or tetrapropylammoniumbromide (TPABr) [51]. In this one-step process, nucleation and growth processes take place in the presence of a support. As a matter of fact, the zeolite layer should be formed from nuclei that appear during the hydrothermal treatment. However, this simple strategy presents some drawbacks. Their number and distribution homogeneity on the support depend on a number of parameters, such as the surface properties, that are difficult to control. Moreover, the formation of nuclei competes with crystal growth processes, which might limit the nuclei density due to mass-transfer limitations. This implies in practice the need to grow thick zeolite films to obtain continuous and well-intergrown layers. This has the advantage of promoting nucleation on the support, avoiding crystal nucleation in the homogeneous phase and reducing the waste reactants. The advantage of this technique is that it allows strict control of the zeolite amount deposited but formation of cracks in the amorphous layer could be a major setback. Cracks formation in the membrane, at times, is induced by the mismatch in the thermal expansion coefficients between the support and the zeolite, causing a drastic reduction of membrane selectivity.

To overcome this limitation, "vapour phase transport (VPT)" technique was proposed [47,48]. VPT is a two-step method that involves: (i) coating a support with amorphous gel containing silica and aluminum; and (ii) followed by crystallization method. With this method strict control of zeolite amount deposited is possible but formation of cracks in the amorphous layer is a set-back. Improvement on membrane synthesis and reproducibility resulted in the development of a secondary growth technique (SGT) [52-55]. In SGT, quality and reproducibility of zeolite membranes can be improved by seeding the support prior to hydrothermal synthesis, decoupling the nucleation and growth steps [52]. Since the nutrient concentration needed for secondary growth is lower than that required for *in situ* hydrothermal synthesis, further nucleation is strongly reduced and almost all crystal growth takes place over the existing crystal seeds. Furthermore, by carefully controlling the seed layer coating the support using a convenient technique, crystallization of undesired zeolite phases can be discouraged, and the rate and direction of crystal growth can be controlled to a certain extent.

In addition, SGT is a two-step method that involves an initial seeding step with previously synthesized zeolite crystals followed by hydrothermal crystallization. In the seeding step, small zeolite seed crystals are deposited on top of the support and after which a film layer of zeolite crystals is grown under hydrothermal crystallization on the pre-seeded support to form a continuous zeolite film layer. Although this method improves on the initial nuclei density but the zeolite crystal orientation in the layer is subject to the seed characteristics, which adversely influence the transport properties of the membrane [56]. Furthermore unavailability of dependable perfect seeding technique to form uniform seed distribution on the surface of support makes application of seeded growth technique impractical to forming reproducible high defect-free zeolite membranes [57].

Application of Thin-Film Zeolite Membranes to Xylene Synthesis

Only few useful studies which deal with the pervaporation (PV) separation of xylene isomers using zeolite membranes have been published in the literature. Table 4 collects the most representative separation data obtained at ambient pressure and near-room temperature. In a series of papers, Nishiyama and co-workers [58-60] have studied the PV separation performance of aromatics within ferrierite (FER) and mordenite (MOR) membranes synthesized by dry gel conversion. The highest selectivities have been reported on FER membranes, achieving values as high as 600 for benzene/p-xylene separation (benzene feed concentration=0.5 mol %) at 303 K and higher than 16 for p-xylene/m-xylene (p/m) separation. However, in all cases, the pervaporated fluxes are very low, showing values lower than 0.01 $\mu mol.m^{-2}.s^{-1}$ in the case of p/m separation. Moreover, PV fluxes were drastically reduced as a result of fouling probably due to the role of nanosized defects and grain boundaries, as pointed out by Wegner et al. [26].

In the case of MFI-type zeolite membranes, the main limitation of PV for p-Xylene separation has been ascribed to the high xylene loadings that are achieved in the MFI structure, especially at near-room temperature. High xylene loadings might cause a distortion of the MFI unit cell (transition from ORTHO to PARA phase) involving channel 'swelling', leading to single-file diffusion. Single-file diffusion of xylene isomers can also be promoted in nanosized grain boundaries between adjacent zeolite grains. In this diffusional regime, xylene isomers with the slowest permeation rate (i.e. o-Xylene and m-Xylene) might limit diffusion, blocking p-Xylene separation and reducing (or even inversing) p-Xylene selectivity. Unlike the separation of lighter gases, energetic and entropic effects play a major role in xylene isomer separation and ultimately determine the membrane performance.

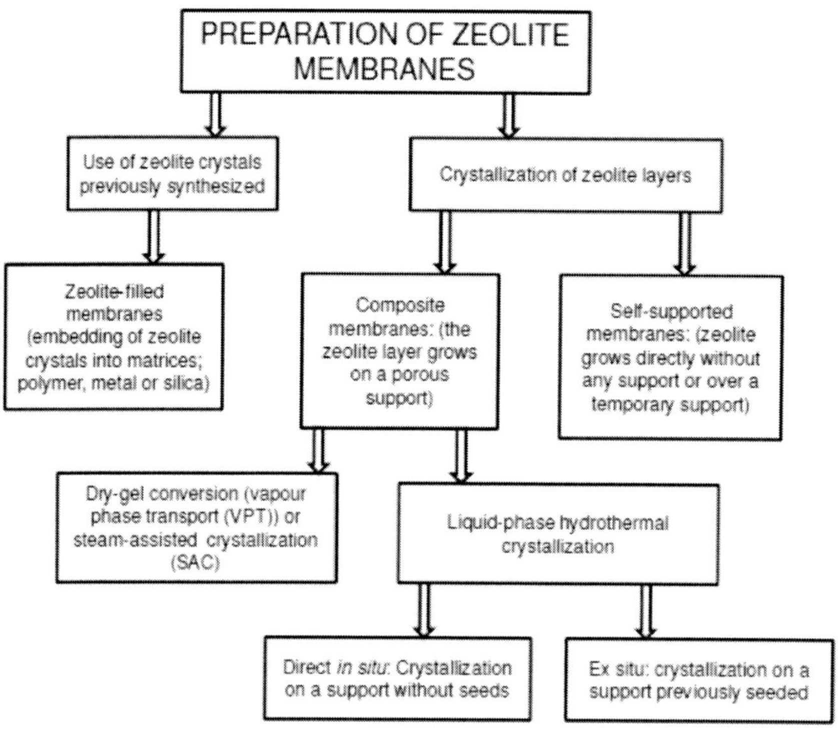

Figure 6. Synthesis strategies for zeolite membranes preparation.

The best p-Xylene separation performance by PV from xylene isomer mixtures has been reported to date by Yuan et al. [61] on template-free hydrothermally synthesized MFI-type zeolite membranes (silicalite-1 membranes), showing p-Xylene fluxes and p-xylene/o-xylene-(p/o)-selectivities up to 13.7 $\mu mol.m^{-2}.s^{-1}$ and 40, respectively. Xiang and Ma [62] have also reported MFI-type zeolite membranes displaying partial selectivity to p-Xylene, but with much higher p-Xylene fluxes (~75 $\mu mol.m^{-2}.s^{-1}$) during PV separation of xylene isomers. Yuan et al. [61] have reported that the room-temperature p-Xylene selectivity of MFI-type zeolite membranes depends strongly on the p-Xylene concentration in the liquid feed. At higher p-Xylene loadings, single-file diffusion is promoted due to an increased distortion of the MFI unit cell, the membrane losing its ability to discriminate p-Xylene among the other xylene isomers. The same authors have also reported extremely long stabilisation periods (>24 h) to achieve steady-state p-Xylene selectivities on the grounds of a slow and progressive distortion of the MFI unit cell upon p-Xylene adsorption and diffusion.

A relatively higher number of studies compared to PV have been published in the past years reporting on the xylene isomer separation performance of MFI membranes by VP. Table 5 collects the most representative results. Unlike PV, at sufficiently low xylene partial pressures (<2 kPa) and in the temperature range 295-673 K, MFI membranes can show optimal selectivity for p-Xylene separation by VP. The permeation performance of these membranes is more dependent on molecular sieving than on preferential adsorption. In particular, Lai et al. [56] and Xomeritakis et al. [63,64] have shown that b-oriented MFI membranes show much better p/o separation factors (up to 480) than in the case of c-oriented or random MFI membranes with a value of 200 $nmol.m^{-2}.s^{-1}.Pa^{-1}$ (p-xylene permeance) at 473 K for a feed mixture of 0.45 and 0.35 kPa p-xylene and o-Xylene partial pressures, respectively.

The evolution of pure p-xylene (PX), m-xylene (MX) and o-Xylene (OX) fluxes with temperature shows a characteristic maximum in the range 523-673 K (denoted as T_{max}) for all the membranes considered in Table 3. The value of the maximum with temperature for p-Xylene flux depends not only on the competition between adsorption and surface diffusion phenomena in the zeolite channels, but also on the amount and characteristics of nanosized defects and grain boundaries in the zeolite matrix. A detailed parametric study on the effect of temperature and feed partial pressure on the xylene permeation and separation performance of MFI-type zeolite membranes has been published by Yeong et al. [65]. In the case of randomly oriented MFI membranes, the highest p/o and p/m separation factors and fluxes at low xylene partial pressures was reported by Gump et al. [66] with self-supported isomorphous BZSM-5 films.

Unlike membrane films, the improved separation performance of these materials should be ascribed to a reduction of intercrystalline defects in the zeolite material due to its intimate confinement in the porous network of the support. Also, the p-Xylene separation capacity of MFI-type zeolite membranes can be tuned to a certain extent by exchanging the cation in the zeolite framework. Tarditi et al. [67] have reported that Ba-exchanged ZSM-5 membranes show much higher p/o and p/m-Xylene separation factors compared to Na and Sr-exchanged counterparts, despite a strong reduction of p-Xylene fluxes probably due to the decrease in the pore size of the channels.

Table 3. Literature survey on xylene isomer separation by VP at low xylene partial pressures (<1 kPa) using Silicalite-1 zeolite membranes (adapted from ref. [5])

Membrane	Support material / geometry	Thickness (μm)	T_{max} (K)	Π_{max}p-Xylene ($nmol.m^{-2}.s^{-1}.Pa^{-1}$)	SF (p/o) (-)	Refs.
b-oriented MFI films[a]	α-alumina / disk	1	423	200	480	[56] [68]
c-oriented MFI films[a]	α-alumina / disk	30	423	30	3	[56] [68]
h0h-oriented MFI films[a]	α-alumina / disk	2	403	40	60	[56] [68]
B-ZSM-5 film[b]	SS / tubular (inner surface)	30	425	2.6	60	[66]
MFI film (self-supported)[b]	Temporary Teflon[R] support	90	473	82	250	[69]
Ultra thin MFI films[a]	α-alumina / disk	0.5	663	300 (663 K) 600 (373 K)	3 16	[70]
Silicalite-1 film[b]	α-alumina / disk	3	400	12	60	[71]
Silicalite-1 film[a]	γ-alumina / tubular (outer surface)	2-10	480	4	1	[72]
Silicalite-1 film[b]	SS/disk	4	373	1220	2.4	[73]
MFI film[b]	α-alumina/ tubular		573	9.5 (523 K)	17.8	[74]
MFI film[a]	α-alumina / disk	1-40	548	20 (398 K)	60-300	[64]
Silicalite-1 film[c]	SS / tubular (outer surface)	25	673	11.9	8	[75]
Al-ZSM-5/Silicalite-1 (bi-layered)[e]	Sintered porous SS/disk	12	373	0.4	~5	[76]
BaZSM-5 (100% Ba^{2+})[f]	SS / tubular	15-20	543	65	19	[67] [77]
Silicalite-1 film[g]	α-alumina / disk	2-3	573	23	~20	[78]

[a]Prepared via seeded hydrothermal synthesis ; [b]Prepared via *in situ* seeded hydrothermal synthesis; [c]Prepared via hydrothermal synthesis; [d]Prepared via pore-plugging *in situ* hydrothermal synthesis; [e]Prepared via *in situ* seeded hydrothermal synthesis (3 cycles); [f]Prepared via secondary growth; [g]Prepared via *in situ* hydrothermal crystallization.

Nanocomposite MFI-Type Zeolite Membranes and Application to Xylene Synthesis

In 2006, Miachon and his co-workers proposed a useful technique (pore-plugging technique) which involves growing zeolite crystals within the pores of support until the pore of the support is completely blocked with the zeolite materials [79-82]. Membranes fabricated through this method are continuous composite zeolite membranes. Advantages of these membranes over conventional TFZ membranes include (i) minimization of cracks formation resulting from thermal expansion mismatch between the support and the zeolite crystals ; (ii) easy to scale-up the preparation procedures at the commercial level as the route is less challenging than he conventional route and; (iii) easiness of membrane handling and module assembling as the separative layer is embedded within the pores of the ceramic support, eliminating the effects of abrasion and shocks.

Excellent xylene isomer separation performance of nanocomposite MFI-type (silicalite-1) membranes at higher xylene loadings (or higher xylene compositions) over their TFZ counterparts has been demonstrated [83-86]. Also, the membrane displayed continuous increase in selectivity at increased xylene loadings/partial pressures [83]. Nanocomposite silicalite-1 zeolite membranes demonstrated by the authors were prepared on α-alumina tubes and hollow fibre supports.

In addition, ternary xylene vapour mixture separation was carried out on the fabricated membranes via vapour permeation (VP) using a typical laboratory VP permeation set-up depicted in Figure 6. Para-Xylene, meta-xylene and ortho-xylene partial pressures in the feed were 0.62, 0.27 and 0.32 kPa, respectively, and the temperature of the membrane system was varied from 423 to 673 K. In the studies, stabilization time of ~ 5 hours was allowed for the the system to attain a steady state before analysis of the streams was done with a gas chromatograph (*Shimadzu GC-14A*), equipped with a solgel-wax capillary column and a flame ionization detector (FID). Also all the tubes were heated and maintained at 393 K to prevent xylene condensation, a major problem in VP. The separation was carried out under Wicke-Kallenbach method (($\Delta P \approx 0$) conditions to avoid viscous flow within the tubes during the separation. Excellent performance of the nanocomposite architecture MFI-type zeolite membranes at higher loadings of xylenes was found to be contradictory to the behaviour of TFZ membranes where loss in selectivity is observed at increasing xylene loadings at higher xylene loadings/vapour pressures (see Figure 7). The outstanding behavior of the nanocomposite MFI-type zeolite membranes during xylene mixture separation was attributed to the nanocomposite architecture of the membranes, preventing distortion of the MFI-zeolite framework at higher concentration of xylene. Thus, the outstanding behaviour of the nanocomposite MFI-type zeolite membranes over TFZ membranes could make them better candidates for separation application in the future development of energy-efficient process for xylene isomer separation based on membrane technology.

Furthermore, the interesting work on the synthesis and application of nanocomposite MFI-type zeolite membrane for xylene isomers separation was later extended to hollow fibre configuration. Although the use of hollow fibre geometry has long been a solution for improving the performance of membrane-based separation processes but industrial application of this type of configuration is limited to water treatment where polymer hollow fibres are commonly used. It has been pointed out in this chapter that polymeric membranes are unsuitable for xylene isomers separation due to their chemical and thermal instability compared to the inorganic/zeolite membranes (refer to Table 3). The concept of nanocomposite architecture was extended to the development of nanocomposite MFI-type zeolite hollow fibre membranes for gas separation. Studies by Alshebani et al. revealed that pure gas permeance of nanocomposite MFI-type (silicalite-1) hollow fibre membranes is about three times higher than the pure gas permeances from nanocomposite Silicalite-1 zeolite tubular membranes. Also, nanocomposite Silicalite-1 zeolite fibers offer higher module surface area/volume ratios (~3000 m^2/m^3) [82].

Based on the interesting results documented by Alshebani et al. [82], separation performance of nanocomposite silicalite-1 hollow fibre membranes during xylene isomer separation was evaluated [84,86]. Nanocomposite silicalite-1 zeolite hollow fibre membranes evaluated for xylene isomer separation via vapour permeation were fabricated on α-alumina supports (o.d. 1.65 mm, i.d. 1.44 mm, porosity 43%) and synthesized using pore-plugging synthesis hydrothermal synthesis technique [79-81]. The fabricated membranes with membrane thickness ~1 μm, were examined under scanning electron microscope (SEM) for morphology. Figure 8 shows a typical SEM image of a nanocomposite Silicalite-1 zeolite membranes revealing presence of zeolite crystals within the pores of the support. Ternary xylene vapour mixture separation was carried out on the fabricated membranes via vapour permeation (VP). Para-Xylene, meta-xylene and ortho-xylene partial pressures in the feed were 0.62, 0.27 and 0.32 kPa, respectively, and the temperature of the membrane system was varied from 423 to 673 K. In the studies, stabilization time of ~ 5 hours was allowed for the the system to attain a steady state before analysis of the streams was done with a gas chromatograph (*Shimadzu GC-14A*), equipped with a solgel-wax capillary column and a flame ionization detector (FID). Also all the tubes were heated and maintained at 393 K to prevent xylene condensation, a major problem in VP. The separation was carried out under Wicke-Kallenbach method (($\Delta P \approx 0$) conditions to avoid viscous flow within the fibres during the separation.

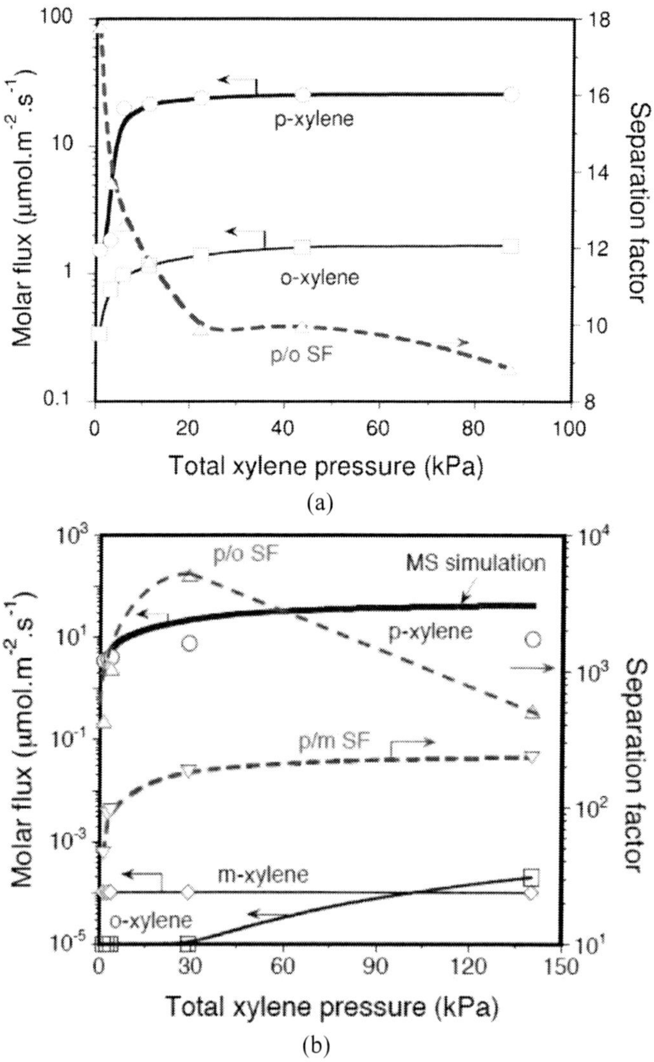

Figure 7. Separation performance of MFI-type (silicalite-1) zeolite membranes at high xylene loadings. (a)TFZ (silicalite-1) membranes (adapted from ref. [74] (b) nanocomposite MFI-type (silicalite-1) zeolite membrane. The straight and dashed lines, respectively, for xylene fluxes and separation factors are a guide to the eye.(obtained with permission from ref. [83]).

The results of VP separation of ternary xylene mixture with the nanocomposite silicalite-1 hollow fibre membranes showed that the membranes were highly selective to p-Xylene at higher temperature and also displayed about 30% increase in p-Xylene flux over an equivalent nanocomposite silicalite-1 membrane tube operated at the same operating conditions (see Figure 9) [84,86]. The authors attributed the enhancement of p-Xylene flux to smaller membrane effective wall thickness (<1 μm) compared to membrane effective thickness of 3 μm for the membrane tube [84]. In addition, it is expected that as a result of the higher membrane surface-to-volume ratios of the hollow fibre membranes compared to more conventional membrane tubes, higher p-xylene permeances are expected leading to higher productivity. For instance, a 5-fibre bundle occupying the volume meant for a single alumina tube could double the p-Xylene productivity operating at the temperature where the maximum p-xylene flux is obtained, taking into account the higher separation surface of the bundle (5 x 9.94 ~ 50 cm^2 vs. 26 cm^2 for alumina tubes) [84]. Therefore, the results described in this section on the application of nanocomposite silicalite-1 zeolite hollow fibre membranes for xylene synthesis open up a research line to scale-up the fibre preparation process to obtain fibre bundles for xylene isomer separation. Using this type of system might be instrumental in enhancing p-xylene productivity and also in reducing both the size and cost of permeating modules for future xylene separation processes.

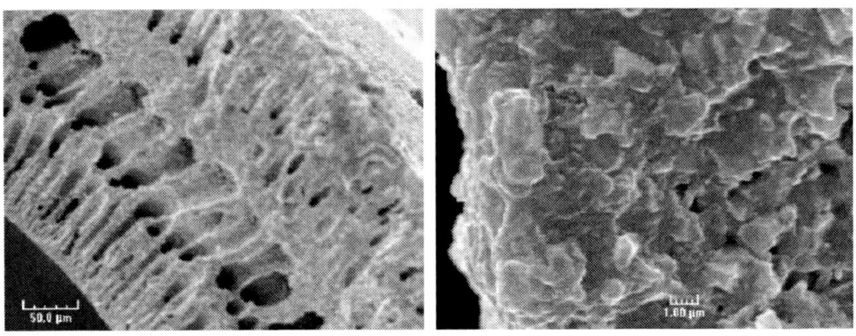

Figure 8. SEM micrographs of the hollow fibre support (left) and fabricated nanocomposite MFI-type hollow fibre membrane (right) (used with permission from ref. [86]).

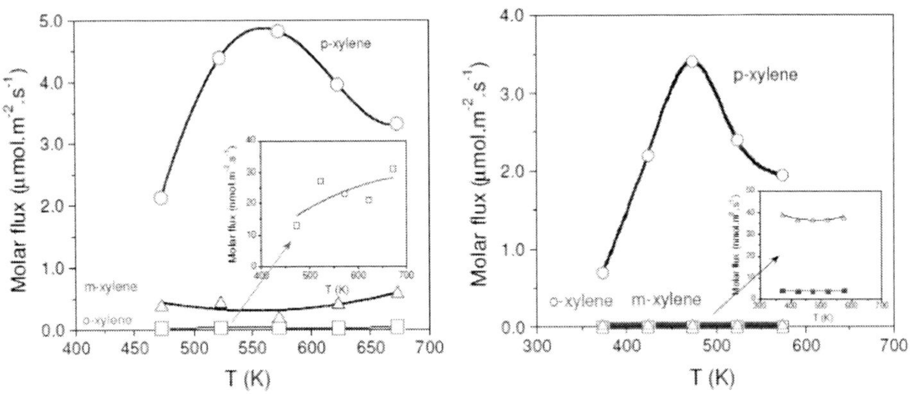

Figure 9. Ternary vapour mixture separation of xylene with nanocomposite MFI-type zeolite membranes as a function of temperature: hollow fibre membrane (left); tubular membrane (right) (obtained with permission from ref. [84]).

Performance of Nanocomposite MFI Membranes during Xylene Isomer Separation and Operating Conditions

Effect of temperature on separation of xylene isomers in MFI-type zeolite membranes (TFZ or nanocomposite) always show the trend depicted in Figure 10. According to Figure 10, p-Xylene transmembrane flux shows a maximum value of about 3.5 $\mu mol.m^{-2}.s^{-1}$ at 473 K, corresponding to a permeance of about 11 $nmol.m^{-2}.s^{-1}.Pa^{-1}$. Also, the p-Xylene flux decreases monotonically after the maximum with no further increase at temperatures higher than 700 K. Maxwell-Stefan (MS) fitting on the experimental data obtained from xylene isomers separation with the nanocomposite MFI-the values showed that the membrane has MFI effective thickness and activation energy for p-Xylene diffusion of $\ell = 0.83 \pm 0.04$ μm and $E_D = 60 \pm 2$ $kJ.mol^{-1}$, respectively [83]. According to the authors, these values were comparable with the value of about 55 $kJ.mol^{-1}$ measured by Masuda et al. [87] on H-ZSM-5 powders using the constant volume method, but significantly higher than the value of about 30 $kJ.mol^{-1}$ measured by Ruthven et al. [88] and Niessen et al. [89]on large silicalite-1 single crystals. In contrary to the work of Grahn et al., in which suggestion regarding alteration of the p-Xylene adsorption pattern due to size effects when tuning from microsized to nanosized particles is suggested, the good prediction level of the MS model for the range of xylene total pressures considered on the nanocomposite MFI-type membranes and also put forward by several authors [64,90], suggests that no relevant structural change of the MFI framework occurs upon p-Xylene adsorption at temperatures higher than 373 K (the 'critical' temperature). The presence of MFI crystals of size <200 nm in a nanocomposite MFI-alumina membrane (the size of the support top layer), together with their strong confinement in the porous alumina network, might compensate for the distortion of the MFI framework upon p-Xylene adsorption, avoiding the phase change traditionally observed for MFI powder at temperatures <400 K.

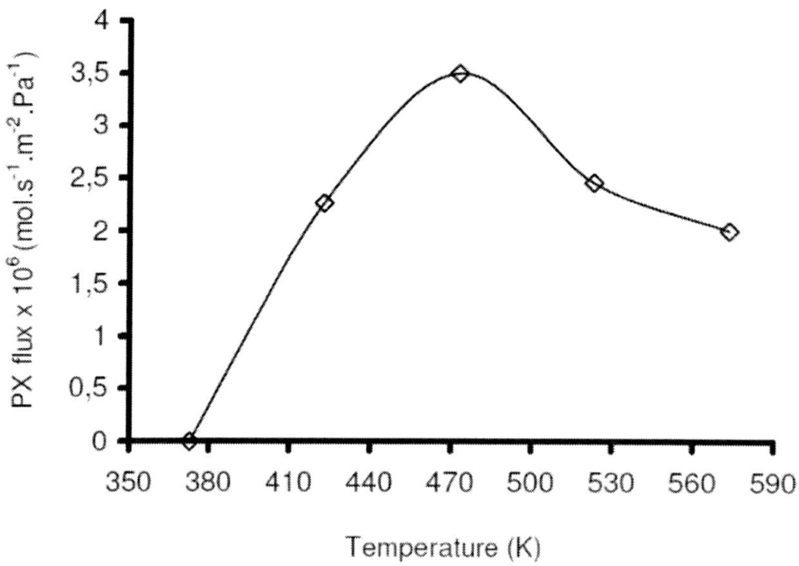

(a)

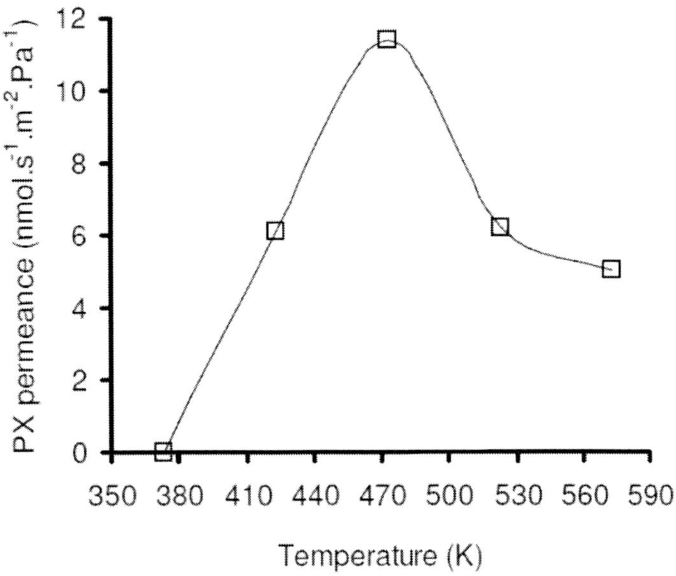

(b)

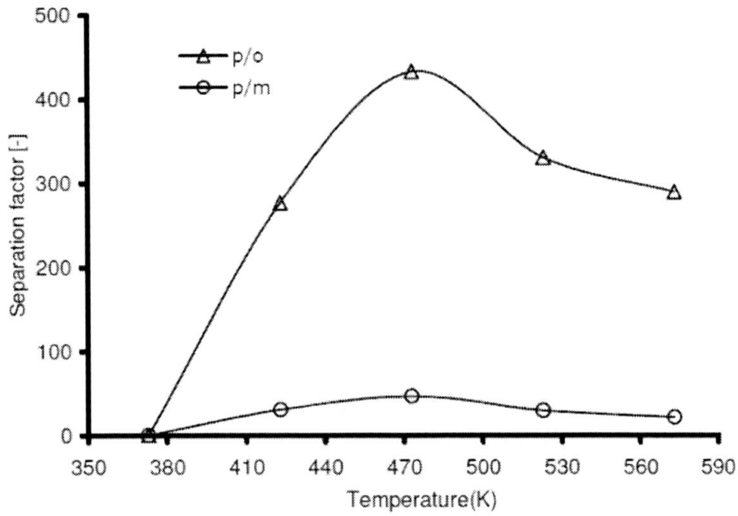

(c)

Figure 10. Performance of nanocomposite MFI-type zeolite membrane during ternary vapour mixture separation of xylene as a function of temperature: P-xylene flux (a); P-xylene permeance (b); membrane separation factor (c) (adapted from ref. [84]).

Furthermore, separation performance behavior of nanocomposite MFI-type zeolite membranes compared to TFZ membranes under the influence of increase in total xylene vapour pressure has been reported [83]. In this study, effect of the total xylene vapour pressure up to 150 kPa on the performance of nanocomposite MFI-type membranes during xylene isomers separation was studied and the p-Xylene flux was found to increase steadily with the total xylene vapour pressure for p/m/o-Xylene ternary mixtures up to 15 kPa and p-Xylene-to-m-Xylene-to-o-Xylene (p/m/o) ratios of 1 : 1 : 3 beyond 30 kPa (see Figure 7).

However, the permeance decreases steadily from 11 to 1 μmol.m^{-2}.s^{-1}.Pa^{-1} but the p/m separation factor remains practically invariable at a value of 200 with the xylene vapour pressure up to 150 kPa. In the case of the p-Xylene/o-Xylene (p/o) separation factor, after showing a maximum value of about 5000 at 30-kPa xylene vapour pressure, it drops drastically to a value lower than 100 at 130 kPa. The increase in p-Xylene flux was attributed to the higher surface coverage of p-Xylene and the retentate/membrane surface, enhancing therefore the p-Xylene driving force across the membrane [83]. At lower xylene partial pressures, keeping the xylene isomers at equimolar composition, selective adsorption of p-Xylene on MFI blocks adsorption of other xylene isomers, thereby paving the way for p-Xylene to permeate. At higher xylene vapour pressure, however, o-Xylene adsorption becomes promoted, showing a slight permeation and contributing therefore to the observed decrease of the p/o xylene separation factor (see Figure 7). This result seems to indicate that, at high xylene coverage, distortion of the MFI framework occurs, higher xylene isomers being transported through the MFI layer. At this juncture, taking into account the similar adsorption properties of the three xylene isomers on MFI zeolites, single-file diffusion can become promoted, the three isomers competing then for passage through the zeolite pores. As a result, the slower permeating isomers (i.e. m-Xylene and o-Xylene) can reduce the permeance of the fastest one (p-Xylene), the membrane selectivity being therefore drastically reduced. Moreover, as put forward by O'Brien-Abraham et al. [91], sorbate-sorbate competition for passage within the MFI pores in xylene mixture vapour permeation might also reduce the accessibility of p-Xylene to MFI channels, hindering therefore its permeation. Unlike film-like membranes (see Figure 7(b)), where the membrane selectivity is strongly affected by the xylene vapour pressure, MFI-type zeolite membranes with nanocomposite architecture have been shown to exhibit high p/o and p/m separation factors for total xylene pressures as high as 150 kPa [83]. The exceptional selective character of the membranes at higher total xylenes pressure was attributed to their nanocomposite architecture, minimizing long term stresses and thus reducing the distortion of the MFI framework at high xylene loadings. Indeed, the high p-Xylene separation performance of the nanocomposite MFI-type zeolite membranes could open up a possible application of these materials to separate xylene isomers via PV. Regarding PV separation of xylene isomers with MFI-type zeolite membranes, only Yuan et al. [61] have reported a p/o selectivity of 60 at 323 K with MFI-type zeolite membranes prepared via templateless seeded hydrothermal synthesis during the separation of an equimolar p/o xylene binary mixture via PV.

In another study carried out on the influence of increase in sweep gas flow rate during VP separation of xylene isomers using MFI-type membranes, increase in the sweep gas flow rate was found to enhance both the p-Xylene flux and membrane selectivity to p-Xylene [83]. For instance, in the work of Daramola et al.[83] p-Xylene flux increased as the N_2 sweep gas flow rate was increased from 5 mL(STP)/min to 30 mL(STP)/min (see Figure 11). The authors ascribed the observation to a reduction of the p-Xylene permeate partial pressure as the sweep gas flow rate increases, contributing to a decrease of the p-Xylene surface coverage at the membrane/permeate surface. The reduction in the p-Xylene surface coverage at the membrane/permeate side leads to an increase in the p-Xylene driving force across the membrane.. the p/m and p/o separation factors show an increase with the sweep gas flow rate, reaching a value as high as 1000 (practically infinite) at 473 K for sweep gas flow rates higher than 30 mL(STP)/min (see Figure 11).

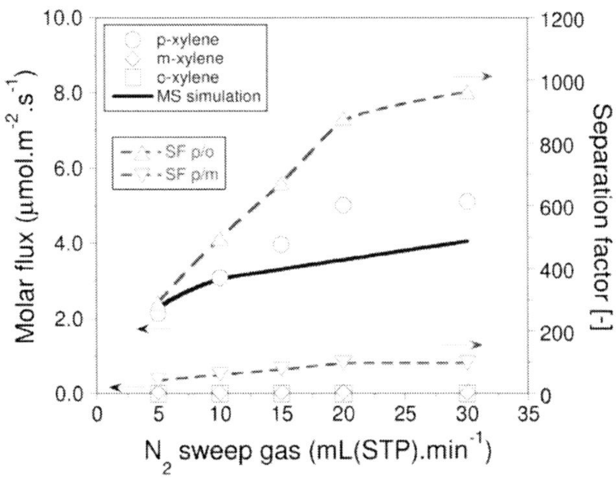

(a)

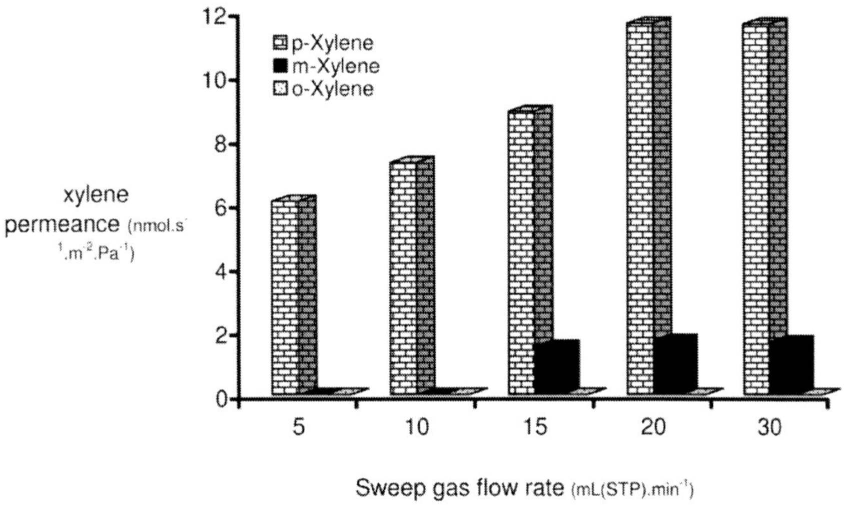

(b)

Figure 11. Performance of nanocomposite MFI-type zeolite membrane during ternary xylene vapour mixture separation as a function of sweep gas flow rate (N_2): xylene flux (a); xylene permeance (b) ((a) obtained with permission from ref. [83], and (b) adapted from [83]).

Support Porous Structure and Performance of Nanocomposite MFI-Type Membranes during Xylene Isomers Separation

Notwithstanding the excellent separation performance of the nanocomposite MFI-type zeolite membranes during xylene isomers separation, low reproducibility the membranes is still a problem [86]. This shortcoming has been ascribed to the intrinsic complexity of the porous structure of hollow fibres (presence of different pore families differing in size and shape) whose nature depends strongly on the manufacturing conditions. As a matter of fact, the strong sensitivity of the pore plugging efficiency in the preparation of MFI-type zeolite membrane tubes on the support pore size has been reported with the maximum admissible value being about 0.5 μm [79]. Extending this conclusion to hollow fibres, the maximum 'effective' pore size of these supports should not exceed, in principle, this critical value. Recently, a study on the role of the porous structure of hollow fibre supports on the separation performance of nanocomposite MFI-type zeolite hollow fibre membranes during xylene isomer separation has been reported [86]. The authors drew correlations between the maximum pore size and /or the form of the pore size distribution of raw hollow fibres on the separation performance of nanocomposite MFI-alumina membranes. In the study, alumina hollow fibres with dimensions listed in Table 4, prepared by a wet spinning process following the methodology described by Goldbach et al. [92], were used to prepare the membranes using pore-plugging hydrothermal synthesis protocol [86].

The results presented in Table 4 suggest an important role of the porous structure of the *raw* hollow fibres (surface pores) on the final membrane quality after synthesis. In principle, hollow fibres displaying maximum pore sizes larger than 0.6 μm, corresponding to First Bubble Point (FBPs) in ethanol lower than 130 kPa, appear to be detrimental to achieving complete pore plugging. This observation, linked to the *n*-butane/H_2 separation factors of the nanocomposite MFI-type zeolite hollow fibre membranes, shows a decreasing trend in the order A→D, being <20 for D-type supports. In the field of membrane-based separation with the application of nanocomposite MFI zeolite membranes, as a reference a *n*-butane/H_2 separation factor > 25 is usually considered as an indicator of good membrane quality in terms of an absence of a large amount of intercrystalline defects [80,86], therefore the results presented for xylene isomers separation also corroborated the results obtained for butane/H_2 test using the same nanocomposite MFI-type hollow fibre membranes (see Table 4). For this separation, the use of A-type supports gave p-Xylene to m-Xylene (p/m) separation factors of at least 50. In all cases, the p/o separation factors remain lower than 15, being slightly promoted for the MFI-type samples prepared on A-type supports. The fact that the hollow fibre membranes prepared here show preferential permeation of o-Xylene rather than m-Xylene in the separation of ternary xylene isomer mixtures was attributed to higher steric constraints of o-Xylene than m-Xylene in the passage within grain boundaries and/or small-sized mesopores, thereby promoting single-file diffusion of the former one [86]. From this study, it is obvious that the mean pore size of the support top layer plays a crucial role on the final membrane quality. The membranes showing the best quality correspond to those prepared with 0.1 and 0.2-μm toplayer supports, being gas-tight before template removal and showing *n*-butane/H_2 separation factor of about 60 after calcinations. In contrast, the membranes synthesized with 0.5 and 0.8-μm support top-layers were not statistically gas-tight before template removal, reflecting an absence of complete pore plugging.

Table 4. Membrane quality of the four hollow-fibre membrane families identified in this study as evaluated from room-temperature n-butane/H_2 and separation and p/m and p/o separation factors at the maximum temperature (range 473-523 K)

Hollow fibre family	Largest pore size (μm)	Separation Factor (SF) n-C_4H_{10}/H_2	Separation Factor (SF) p/m	Separation Factor (SF) p/o
A	0.3-0.5	>300	70 ± 25	9 ± 1
B	0.5-0.6	50-300	19 ± 5	7 ± 1
C	0.5-0.6	20-50	12 ± 5	7 ± 1
D	>0.6	<20	<7	<3

CONCLUSION

As described in this chapter, separation of xylene isomers with membrane-based process, especially MFI-type zeolite membranes, is explorable and feasible. As a matter of fact, nanocomposite MFI-type (silicalite-1) membrane tubes and hollow fibre membranes have been demonstrated to be selective to para-xylene from xylene isomers. Also, it has been demonstrated that membrane configuration and effective membrane wall thickness play a prominent role in enhancing cross membrane flux. Research results reviewed and documented in the chapter reveal that nanocomposite silicalite-1 hollow fibre membrane could enhance para-xylene fluxes during the separation of ternary vapour mixture of xylene isomers due to the smaller effective wall thickness of the membrane (membrane thickness <1 μm) when compared to conventional nanocomposite MFI-type zeolite membrane tubes (membrane thickness >3 μm) [84]. Additionally, hollow fibres offer the added advantage of membrane surface-to-volume ratios as high as 3000 m^2/m^3 compared to conventional membrane tubes [84]. Using this type of system could be instrumental in reducing both the size and cost of permeating modules for future xylene separation processes. However, obtaining high quality nanocomposite silicalite-1 membranes is subject to the availability of high quality membrane supports (fibres or tubes).

Influence of operating parameters like temperature, xylenes pressure, sweep gas flow rate (in case of vapour permeation) on the separation performance of nanocomposite MFI-type zeolite membrane has shown that excellent separation of para-xylene from its isomers could be achieved by minimizing the xylene adsorption level by operating either (i) at lower module temperature (at most 473K) or (ii) at high para-xylene vapour pressure [83]. Also operating at high sweep gas velocity could enhance the separation by reducing the para-xylene vapour pressure in the permeate side, and thus increasing the para-xylene permeance through the membrane [83]. Regardng the xylene isomers separation performance of nanocomposite MFI-type zeolite membranes at higher xylene loadings, the excellent separation performance of these membranes at higher xylene loading (or vapour pressure) could make MFI-type zeolite membranes with nanocomposite architecture a promising choice for pervaporation applications where concentration profile is usually a major problem at higher loading of xylene. However, to make this process economically competitive with existing processes, especially in term of energy consumption, it is advantageous to operate at high vapour pressure because of the increased in para-xylene flux. Also the results reviewed in this chapter have demonstrated that economic viability of membrane separation process rests on excellent membrane flux and availability of defect-free membranes.

Despite the extensive research on the synthesis of selective zeolite membranes, reproducibility remains a major problem. Although zeolite membranes have displayed very high separation performance than those of pure polymeric membranes but costs of ceramic support materials for zeolite-based membranes are very high, contributing to reluctance in accepting the technology industrially. Zeolite membranes are thermally stable with good chemical resistance and mechanical strength in comparison to pure polymer membranes but they are very fragile and brittle. Furthermore, technology for fabricating commercial zeolite membranes is still in the developmental stage while technology for commercial production of polymeric membranes is very mature with applications in series of industrial processes. To address the issue of poor reproducibility, loss of mechanical strength in polymeric membranes and huge costs of production of zeolite membranes, mixed matrix membranes have been proposed.

ACKNOWLEDGMENT

The author acknowledges Elsevier publisher BV, The Netherlands and Taylor and Francis Publisher, USA, for the permission to use some figures from their published work in this chapter.

REFERENCES

[1] Franck, H-G; Stadelhofer, JW. *Industrial Aromatic Chemistry*. Raw Materials, Process, Products, Springer-Verlag, Berlin, 1988.

[2] Xylenes, *Nexant Chem Systems*. May, 2002.(http://nexant.ecnext.com/coms2/)

[3] Fabri, J; Graeser, H; Simo, TA. Xylenes, In: Ullmann's Encyclopedia of Industrial Chemistry, seventh ed., John Wiley & Sons, Inc. Germany, 2009.

[4] Cannella, WJ. in: *Kirk-Othmer Encylopedia of Chemical Technology*, vol. 16, New Jersey, 2001, 841.

[5] Daramola, MO; Burger, AJ; Pera-Titus, M; Giroir-Fendler, A; Miachon, S; Dalmon, JA; Lorenzen, L. *Asia-Pacific Journal of Chemical Engineering*, 5(6), (2010), 815-837.

[6] Laurich, SA. US Patent 3,467,724, *Chevron Research Corporation*, 1969.

[7] Vickers, GH. US Patent 20040010170, *BP American Inc.*, 2004.

[8] Leflaive, P; Barthelet, K. US Patent 200704443253, 2007.

[9] Roesler, CM; Kulprathipanja, S; Rekoske, JE. US Patent 200220103408, 2002

[10] Bessarabov, D. *Membr. Technol.*, 1999, (1999), 9–13.

[11] Spillman, R. *Economics of gas separation membrane processes*, in R.D. Noble and S.A. Stern, eds., Membrane Separation Technology: Principles and Applications, Elsevier, 1995.

[12] Matsuura, T. *Synthetic Membrane and Membrane Separation Processes*, CRC Press, 1993.

[13] Dhingra, SS. Mixed gas transport study through polymeric membranes: a novel technique, PhD dissertation, Virginia Polytechnic Institute, 1997.

[14] Ettouney, HM; El-Dessouky, HT; Waar, WA. *J. Membr. Sci.*, 148, (1998), 105–117.

[15] Henley, E; Santos, M. *AIChE J.*, 1967, 13, 1117.

[16] Ito, A; Hwang, S. *J. Appl. Polym. Sci.* 1989, 38, 483.

[17] Sridhar, S; Khan, A. *J. Membr. Sci.* 1999,159, 209-219.

[18] Smitha, B; Sridhar, S; Khan, AA. *J. Membr. Sci.* 2005, 259, 10-26.

[19] Powell, CE; Qiao, GG. *J. Membr. Sci.*, 2006, 279, 1-49.

[20] Shao, P; Huang, RYM. *J. Membr. Sci.* 2007, 287, 162-169.

[21] Semenova, S1; *J. Memb. Sci.*, 2004, 231, 189-207.

[22] Okamoto, K; Noborio, K; Hao, J; Tanaka, K; Kita, H. *J. Membr. Sci.* 1997, 134, 171-179.

[23] Shimazu, A; Hacchisuka, H; Ikeda, K. *Kagaku Kogaku*, 1996, 60(4), 262-263.

[24] Tsubouchi, K; Yoshikawa, M. Membrane (Maku), 1998, 23(6), 322-326.
[25] Fang, J; Tanaka, K; Kita, H; Okamoto, K. *Polymer*, 1999, 40, 3051-3059.
[26] Wegner, K; Dong, J; Lin, YS. *J. Membr. Sci.* 158(1-2), (1999), 17-27.
[27] Cunha, VS; Paredes, MLL; Borges, CP; Habert, AC; Nobrega, R. *J. Membr. Sci.* 2002, 206, 277-299.
[28] Chen, HL; Wu, LG; Tan, J; Lhu, CL. *Chem. Eng. J.* 2000, 78, 159-164; K. Wegner, J. Dong, Y.S. Lin, J. Membr. Sci., 1999, 158, 17-27.
[29] Hao, J; Tanaka, K; Kita, H; Okamoto, K. *J. Membr. Sci.* 1997, 132, 97-108.
[30] Schleiffelder, M; Staudt-Cickel, C. *Reactive & Functional Polymers*, 2001, 49, 205-213.
[31] Weitkamp, J. *Solid State Ionics*, 131, (2000), 175-188.
[32] Argauer, RJ; Kensington, M; Landolt, GR. *US Patent*, 3, 702, 886, 1972.
[33] Kokotailo, GT; Lawton, SL; Olson, DH; Meier, WM. *Nature*, 272, (1978), 437-438.
[34] Vezzalini, G; Quartieri, S; Galli, E; Alberti, A; Cruciani, G; Kvick, A. *Zeolites*, 19, (1997), 323-325.
[35] Grose, RW; Flanigen, EM. Crystalline silica, *US Patent*, 4, 061, 724, Union Carbide Corporation, New York, 1977.
[36] Flanigen, EM; Bennett, JM; Grose, RW; Cohen, JP; Patton, RL; Kirchner, RM. *Nature*, 271, (1978), 512-516.
[37] Breck, DW. Zeolite molecular sieves: Structure, chemistry and use, John Wiley & Sons, New York, 1974.
[38] Baerlocher, C; Meier, WM; Olson, DH. *Atlas of Zeolite Framework Types*, Fifth ed., Elsevier, Amsterdam, 2001.
[39] Olson, DH; Kokotailo, GT; Lawton, SL; Meier, WM. *The Journal of Physical Chemistry*, 85, (1981), 2238-2243.
[40] Caro, J; Noack, M. *Micropor. Mesopor. Mater.* 115(3), (2008), 215-233.
[41] Julbe, A. Zeolite membranes-synthesis, characterization and application, in: J. Cejka, H. van Bekkum, A. Corma and F. Schuth (Eds.), Introduction to zeolite Science and Practice, *Stud. Surf. Sci. Catal. Elsevier B.V.*, Amsterdam, 2007, 181-219.
[42] Bowen, TC; Noble, RD; Falconer, JL. *J. Membr. Sci.* 245, (2004), 1-33.
[43] Fong, YY; Abdullah, AZ; Ahmad, AL; Bhatia, S. *Chem. Eng. J.* 139(1), (2008), 172-193.
[44] Morigami, Y; Kondo, M; Abe, J; Kita, H; Okamoto, K. *Sep. Purif. Technol.* 25, (2001), 251-260.
[45] Tennison, S. *Membr. Technol.* 128, (2000), 4-9.
[46] Xu, W; Dong, J; Li, J; Wu, F. *J. Chem. Soc.*, Chem. Commun. 10, (1990), 712.
[47] Matsufuji, T; Nishiyama, N; Matsukata, M; Ueyama, K. *J. Membr. Sci.*, 178, (2000), 25-34.
[48] Bonilla, G; Vlachos, DG; Tsapatsis, M. *Micropor. Mesopor. Mater.* 42, (2001), 191.
[49] Matsukata, M; Nishiyama, N; Ueyama, K. *Chem. Commun.* 3, (1994), 339-340.
[50] Matsukata, M; Nishiyama, N; Ueyama, K. *Micropor. Mater.* 1(3), (1993), 219-222.
[51] Davis, BH; Singh, KSW. in: F. Schüth, K.S.W. Singh, J. Weitkamp (Eds.), *Handbook of porous solids*, Vol. 1, Wiley-VCH, Weinheim, 2002.
[52] Boudreau, LC; Kuck, JA; Tsapatsis, M. *J. Membr. Sci.* 152, (1999), 41-59.
[53] Chau, JLH; Tellez, C; Yeung, KL; Ho, K. *J. Membr. Sci.* 164(1-2), (2000), 257-275.
[54] Gouzinis. A; Tsapatsis, M. *Chem, Mater.* 10, (1998), 2497.

[55] Hedlund. J; Noack, M; Kolsch, P; Creaser, D; Caro, J; Sterte, J. *J. Membr. Sci.* 159, (1999), 263.
[56] Lai, Z; Bonilla, G; Diaz, I; Nery, JG; Sujaoti, K; Amat, MA; Kokkoli, E; Terasaki, O; Thompson, RW; Tsapatsis, M; Vlachos, DG. *Science*, 300, (2003), 456-460.
[57] Ge, Q; Wang, Z; Yan, Y. *J. Am. Chem. Soc.* 13, (2009), 17056-17057.
[58] Nishiyama, N; Ueyama, K; Matsukata, M. *Micropor. Mater.* 7, (1996), 299-308.
[59] Nishiyama, N; Matsufuji, T; Ueyama, K; Matsukata, M. *Micropor. Mater.* 12, (1997), 293-303.
[60] Matsufuji, T; Nakagawa, S; Nishiyama, N; Matsukata, M; Ueyama, K. *Micropor. Mesopor. Mater.* 38, (2000), 43-50.
[61] Yuan, W; Lin, YS; Yang, W. *J. Am. Chem. Soc.* 126, (2004), 4776-4777.
[62] Xiang, S; Ma, YH. in: Proceedings of the third International Conference on Inorganic Membranes, Formation and Characterization of Zeolite Membranes from Sols., Worcester, MA, 1994.
[63] Xomeritakis, G; Tsapatsis, M. *Chem. Mater.* 11(4), (1999), 875-878.
[64] Xomeritakis, G; Nair, S; Tsapatsis, M. *Micropor. Mesopor. Mater.* 38, (2000), 61-73.
[65] Yeong, YF; Abdullah, AZ; Ahmad, AL; Bhatia, S. *J. Membr. Sci.* 341(1-2), (2009), 96-108.
[66] Gump, CJ; Tuan, VA; Noble, RD; Falconer, JL. *Ind. Eng. Chem. Res.* 40, (2001), 565-577.
[67] Tarditi, AM; Lombardo, EA. *Sep. Purif. Technol.* 61, (2008), 136-147.
[68] Xomeritakis, G; Lai, Z; Tsapatsis, M. *Ind. Eng. Chem. Res.* 40, (2001), 544-552.
[69] Sakai, H; Tomita, T; Takahashi, T. *Sep. Purif. Technol.* 25, (2001), 297-306.
[70] Hedlund, J; Sterte, J; Anthonis, M; Bons, AJ; Carstensen, B; Corcoran, N; Cox, D; Deckman, H; De Gijnsr, W; de Moor, P-P; Lai, F; McHenry, J; Mortier, W; Reinoso, J; Peters, J. *Micropor. Mesopor. Mater.* 52, (2002), 179-189.
[71] Keizer, K; Burggraaf, AJ; Vroon, ZAEP; Verweij, H. *J. Membr. Sci.* 147, (1998), 159-172.
[72] Baetsch, CD; Funke, HH; Falconer, JL; Noble, RD. *J. Phys. Chem.* 100, (1996), 7676-7679.
[73] Haag, S; Hanebuth, SM; Mabande, GTP; Avhale, A; Schwieger, W; Dittmeyer, R. *Micropor. Mesopor. Mater.* 96, (2006), 168-176.
[74] Gu, X; Dong, J; Nenoff, TM; Ozokwelu, DE. *J. Membr. Sci.* 280, (2006), 624-633.
[75] Tarditi, AM; Irusta, S; Lombardo, EA. *Chem. Eng. J.* 122(3), (2006), 167-174.
[76] Mabande, GTP; Noack, M; Avhale, A; Kolsch, P; Georgi, G; Schwieger, W.; Caro, J. *Micropor. Mesopor. Mater.* 98(1-3), (2007), 55-61.
[77] Tarditi, AM; Lombardo, EA; Avila, AM. *Ind. Eng. Chem. Res.* 47, (2008), 2377-2385.
[78] Zhang, C; Hong, Z; Gu, X; Zhong, Z; Jin, W; Xu, N. *Ind. Eng. Chem. Res.* 48, (2009), 4293-4299.
[79] Miachon, S; Landrivon, E; Aouine, M; Sun, Y; Kumakiri, I; Li, Y; Pachtova Prokopova, O; Guilhaume, N; Giroir-Fendler, A; Mozzanega, H; Dalmon, J-A. *J. Membr. Sci.* 281, (2006), 228-238.
[80] Miachon, S; Ciavarella, P; van Dyk, L; Kumakiri, I; Fiaty, K; Schuurman, Y; Dalmon, J-A. *J. Membr. Sci.* 298, (2007), 71-79.
[81] Li, Y; Pera-Titus, M; Xiong, G; Yang, W; Landrivon, E; Miachon, S; Dalmon, J-A. *J. Membr. Sci.* 325, (2008), 973-981.

[82] Alshebani, A; Pera-Titus, M; Landrivon, E; Schiestel, T; Miachon, S; Dalmon, J-A. *Micropor. Mesopor. Mater.* 115, (2008), 197-205.
[83] Daramola, MO; Burger, AJ; Pera-Titus, M; Giroir-Fendler, A; Lorenzen, L; Dalmon, J-A. *Sep. Sci. Technol.*, 45(1), (2010), 21-27.
[84] Daramola, MO; Burger, AJ; Pera-Titus, M; Giroir-Fendler, A; Lorenzen, L; Miachon, S; Dalmon, J-A. *J. Membr. Sci.*, 337, (2009), 106-112.
[85] Van Dyk, L; Lorenzen, L; Miachon, S; Dalmon, J-A. *Catal. Today*, 104, (2005), 274-280.
[86] Deng, Z; Nicolas, C-H; Daramola, MO; Sublet, J; Schiestel, Th; Burger, AJ; Guo, Y; Giroir-Fendler, A; Pera-Titus, M. *J. Membr. Sci.* 364, (2010), 1-8.
[87] Masuda, T; Fujikata, Y; Nishida, T; Hashimoto, K. *Micropor. Mesopor. Mater.* 23, (1998), 157-167.
[88] Ruthven, DM; Eic, M; Richard, E. *Zeolites*, 11, (1991), 647-653.
[89] Niessen, W; Karge, HG. *Micropor. Mater.* 1, (1993), 1-8.
[90] Talu, O; Guo, CJ; Hayhurst, DT. *J. Phys. Chem.* 93, (1989), 7294-7298.
[91] O'Brien-Abraham, J; Kanezashi, M; Lin, YS. *J. Membr. Sci.* 320, (2008), 505-513.
[92] Goldbach, A; Mauer, T; Stroh, N. *Keram. Z.* 53(11), (2001), 1012-1016.
[93] Vu, DQ; Koros, WJ; Miller, SJ. *J. Memb. Sci.* 211, (2003), 311-334.

In: Xylenes
Editor: Michael Olawale Daramola

ISBN: 978-1-62808-342-2

Chapter 6

NANOCOMPOSITE MFI-TYPE ZEOLITE CATALYTIC MEMBRANE REACTORS FOR ISOMERIZATION OF M-XYLENE TO P-XYLENE

***M. O. Daramola*[1]*, A. J. Burger*[2] *and A. Giroir-Fendler*[3]**

[1]School of Chemical and Metallurgical Engineering, Faculty of Engineering & the Built Environment, University of the Witwatersrand, Johannesburg, South Africa
[2]Department of Process Engineering, University of Stellenbosch, Stellenbosch, South Africa
[3]Engineering Research Group, Institute for Research on Catalysis & Environment, Villeurbanne Cedex, Lyon, France

ABSTRACT

Future production of chemicals (e.g. fine and specialty chemicals) in industries is faced with a challenge of limited material and energy resources. However, process intensification might play a significant role to alleviate this problem. Vision of process intensification through multifunctional reactors has stimulated research on membrane-based reactive separation processes, in which membrane separation and catalytic reaction occur simultaneously in one unit. These processes are rather attractive applications because they are potentially compact, less capital intensive, and have lower processing costs than traditional processes. Therefore in this chapter, a short overview of potential application of membrane technology in the production of xylene, in particular application of zeolite-based membranes as extractor-type catalytic membranes in isomerisation of less used xylenes to p-Xylene is presented.

Keywords: Xylenes, membrane reactors, process intensification, membranes, isomerization

INTRODUCTION

Mixed xylenes (from the Greek *xylon*=wood), first discovered in crude wood spirit in 1850 by Cahours, constitutes a family of C_8-aromatics with molecular formula C_8H_{10} including three constitutional isomers: o-Xylene (OX), m-Xylene (MX) and p-Xylene (PX). These constitutional isomers are referred to as mixed xylenes. The xylene isomers differ from one another by the relative position of the two methyl groups in the benzene ring. The most industrially used xylene isomer is para-xylene. Para-xylene is almost exclusively used as raw material in the production of terephthalic acid (TPA) and dimethyl terephthalate (DMT), which are reacted with ethyleneglycol to form polyethylene terephthalate (PET), the raw material for polyester resin. Polyester resin in the manufacture of polyester fibres, films and fabricated items (e.g. beverage bottles).

Production of p-Xylene via isomerization is a chemical-equilibrium restricted reaction process [1]. At equilibrium, the amount of o-Xylene increases with increasing temperature while amount of p-Xylene seems to be independent of temperature at higher temperatures (see Figure 1). Also, m-xylene conversion at equilibrium is about 50% in conventional Fixed-bed reactors (FBRs). Therefore, realization of total conversion during xylene isomerization process in conventional catalytic reactors (fixed-bed reactors) is impossible. Thus, existing industrial technology could only produce equilibrium or near equilibrium xylene mixtures. Recycling the xylene streams back into the process lines might ensure higher p-Xylene productivity, but at the expense of higher operational costs due to higher energy consumption. However, the use of extractor-type catalytic membrane reactors (e-CMRs) has been demonstrated to possess the potential to eliminate equilibrium restriction associated with equilibrium-limited reactions in fixed-bed reactors with a drastic reduction in operational costs resulting from a reduction in energy consumption. The enormous potential of large-scale applications of xylene isomerization in oil and petrochemical industries promises major advances and development of such systems in a near future. However the development of such systems for high purity p-Xylene from xylene isomerization is retarded due partly to inadequate understanding of the system and to the absence of high-flux and defect-free membranes that have high selectivity for p-Xylene, especially at industrial operating conditions. To overcome this obstacle, in-depth understanding of the fundamental behaviour of the system and availability of high-flux membranes, having high selectivity for p-Xylene, is essential. In this chapter, an overview of the application of zeolite catalytic membrane reactors for xylene synthesis is presented. In particular, the chapter aims at showcasing the potentials of MFI-type zeolite catalytic membrane reactors for xylene synthesis, especially, the conversion of less used xylene isomers (ortho-xylene and meta-xylene) to para-xylene.

CONVENTIONAL REACTORS AND ZEOLITE CATALYTIC MEMBRANE REACTORS

Over the next 25 years, worldwide industrial energy consumption is expected to grow from 183.8 quintillion Joule (J) in 2006 to 257.9 quintillion Joule (J) in 2030 at an average annual rate of 1.4% [2]. This trend is attributable to the use of energy-intensive conventional processes for chemical conversion and separation in chemical industries. Conventional process for synthesis of chemicals in chemical consists of a reaction unit followed by a separation unit (see Figure 2). In the reaction unit, conversion of reactants to desired and undesired products occurs and removal of desired product from the reaction mixture takes place in the separation unit. Through this conventional system, huge energy is consumed translating into huge operating costs. Therefore, it is expected that introduction of more energy-efficient technologies in the chemical industry could contribute significantly to nationwide and worldwide energy savings.

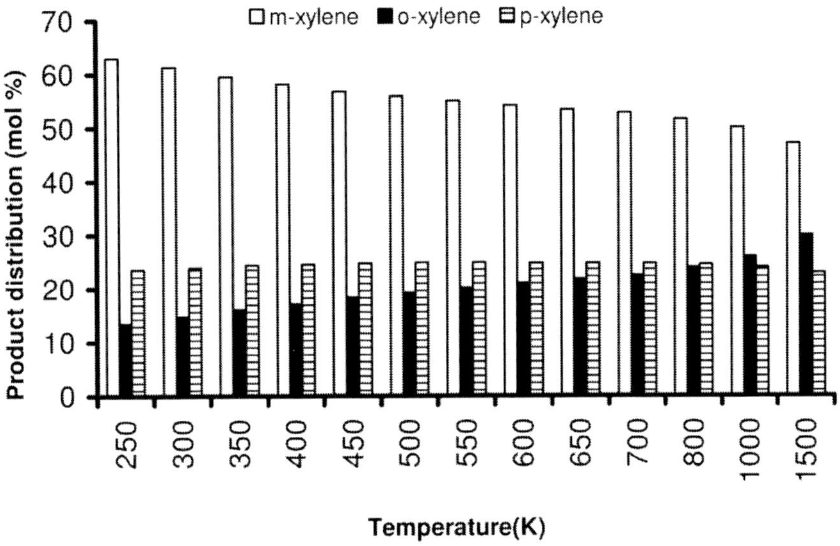

Figure 1. Equilibrium product distribution (molar basis) as a function of temperature in the range 250-1500 K. (Adapted from ref.[1]).

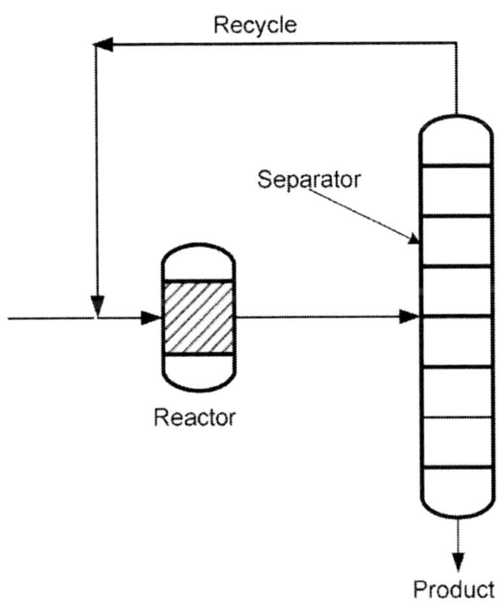

Figure 2. Conventional reaction followed by separation in chemical industry (adapted from refs. [3,4]).

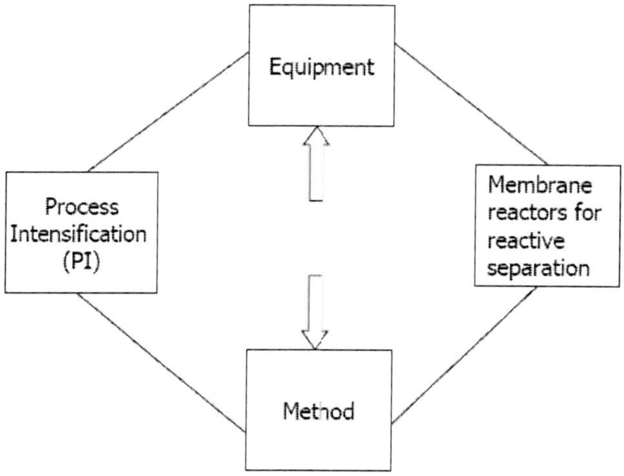

Figure 3. Process intensification and its components for membrane reactor concept for reaction coupling separation process (adapted from ref. [4]).

Reduction in energy consumption in chemical industry could be achieved through process intensification of chemical processes with the incorporation of knowledge acquired from membrane science and technology. Combination of membrane-based separation science and technology and reaction/catalysis engineering could result into an intensified unit referred to as catalytic membrane reactors (CMRs). A concept of process intensification for chemical industry involving the use of membrane reactors is depicted in Figure 3.

According to the IUPAC definition, a membrane reactor (MR) is a device combining a membrane-based separation and a chemical reaction in one unit [5]. Also, MR is a device for carrying out a reaction and a membrane-based separation simultaneously in the same physical enclosure or in close proximity [6]. Meanwhile, a membrane is defined essentially as a barrier, which separates two phases and restricts transport of various chemicals in a selective manner. Concept of membrane reactor for reaction coupling separation is depicted in Figure 4. In the concept, membranes are applied for selective removal of the target product from the reaction zone.

Zeolite-Based Catalytic Membrane Reactors

Zeolite Catalytic Membrane Reactors (ZCMRs), where zeolite membrane separation is coupled with catalytic reaction in the same unit, are attractive applications because they have been demonstrated to be potentially compact, less capital intensive and have lower operating costs than more conventional processes. Two main types of zeolite membrane reactors have been identified namely: inert zeolite catalytic membrane reactors (IZCMRs) and the zeolite catalytic membrane reactors (ZCMRs). In the IZCMRs, the membrane is not catalytically active and does not participate in the reaction but it simply acts as selective separation unit for the desired products while acting as a barrier to the reactants and undesired products. Inert zeolite catalytic membrane reactor allows catalyst pellets to flow with the reactants on the feed side (usually the inside of the membrane). This type of configuration is also referred to as Inert Membrane Reactor with Catalyst on the Feed side (IMRCF). Schematic of IZCMR is depicted in Figure 5. On the other hand, a zeolite catalytic membrane reactor (ZCMR) has a zeolite membrane that has either been coated with or is made of a material that contains catalyst, indicating that the membrane itself participates in the reaction. Some of the reaction products (those that are small enough) pass through the membrane and exit the reactor on the permeate side [7].

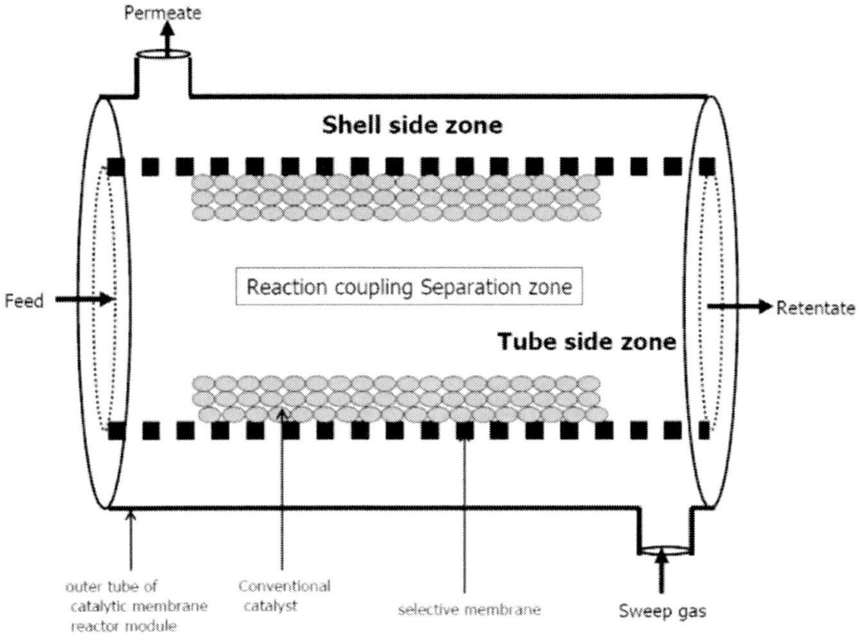

Figure 4. Schematic of a typical membrane reactor (Packed-bed membrane reactor) for reaction coupling separation process.

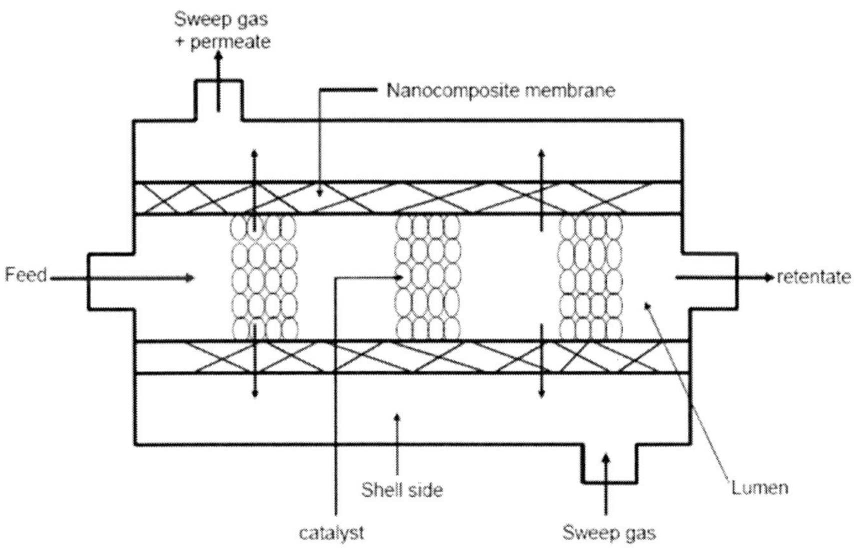

Figure 5. Schematic of inert zeolite catalytic membrane reactor (IZCMR) with catalyst packed inside the lumen of zeolite membrane

On the basis of the way zeolite membrane and the catalyst are combined, ZCMRs have been broadly classified as: (i) extractor-type (e-ZCMR), (ii) distributor-type (d-ZCMR), and (iii) contactor-type (c-ZCMR [8]. Schematically, Figure 6 compares the three ZCMR configurations. In all cases, the membrane can show inherent catalytic character or only act as a separation/contactor unit between the phases and the catalyst. More specific details about these configurations and application domains can be found in literature [9-11].

Among the types of ZCMRs, extractor-type ZCMRs (e-ZCMRs) are by far the most widespread application of CMRs. Classical applications of this configuration range from dehydrogenation, isomerization and esterification/etherification reactions to hormone synthesis and wastewater biological treatment. In this configuration, selective removal of one/more products from the reaction zone enhances the conversion of the reaction by shifting the equilibrium position or by promoting the catalytic activity. However, to overcome the equilibrium restriction, the reaction must be sufficiently fast compared with the mass transport through the membrane (kinetic compatibility). A special benefit can be that the removal of one of the products provides an integrated product purification thus decreasing the number of process units. Also an activity improvement is possible through selective removal of reaction rate inhibitors. As a result of the need for more energy-efficient technologies in the chemical industry, research efforts on the development of e-ZCMRs have been intensified. Analysis of ZCMRs is based on the fundamental principles usually employed for conventional reactors (e.g. plug flow reactors), except transport law accounting for the transport of molecule across the membrane that is included in the mathematical expression. Basically, analysis of ZCMRs involves:

- Mole balance in the catakytic bed : Material in –Material out + Generation =Accumulation
- Rate law that accounts for disapperance of reactant: $R = kC^n$. Where R, the reaction rate; k; reaction rate constant; C, reactant concentration and n, the reaction order
- Transport law accounting for the transport or flux of product through the membrane: $J = k_m \Delta C$. Where J, the flux of the product through the membrane; k_m, the mass transfer coefficient; and ΔC, the concentration gradient across the membrane. It is noteworthy to mention that the transport law takes into account adsorption-diffusion mechanism that governs the transport of molecule through zeolite membranes

Extractor-Type MFI-Type Zeolite Catalytic Membrane Reactors (e-ZCMRs) for Xylene Synthesis

The use of IZCMRs having MFI-type zeolite membranes as seperative layers has been proposed by several authors to promote para-xylene recovery during para-xylene synthesis. In this concept, selective removal of p-Xylene by the MFI-type zeolite membranes could promote the p-Xylene yield and productivity when compared to convectional fixed-bed reactors (FBRs). The scheme of this concept is represented schematically in Figure 7.

Between 2005 and 2012, about eighteen (18) articles have been published in open scientific journals and conferences on the potential application of e-ZCMR equipped with

MFI-type zeolite membrane in xylene synthesis (see Figure 8). The development of e-ZCMRs compared to FBRs is not at a good progression due to difficulty associated with fabricating and reproducing high quality and highly p-Xylene-selective MFI-type zeolite membranes for e-ZCMRs application.

In the literature, two types of extractor-type catalytic membrane reactors namely inert zeolite catalytic membrane reactor (IZCMRs) and catalytically active zeolite catalytic membrane reactors (CAZCMRs) have been studied for p-Xylene synthesis. In the IZCMR, the MFI-type zeolite membrane acts as a non-catalytic p-Xylene extractor while in the CAZCMR, the MFI-type zeolite membrane displays catalytic activity for xylene isomerization as well as p-xylene extractor. Figure 9 illustrates different configurations of MFI-type zeolite membrane-to-catalyst combinations in e-ZCMRs for xylene isomerization [14].

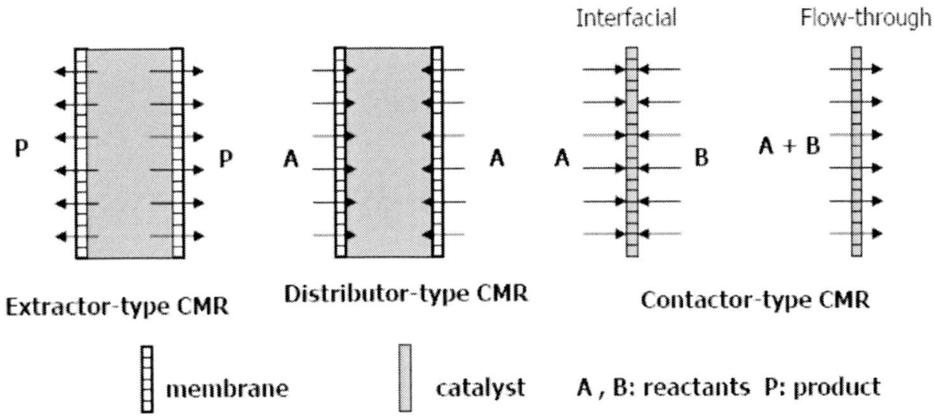

Figure 6. Classification of CMRs: (a) extractor, (b) distributor, (c) flow-through contactor and (d) interfacial contactor. A and B represent reactants while P, P1, P2 are the products (adapted from ref.[4]).

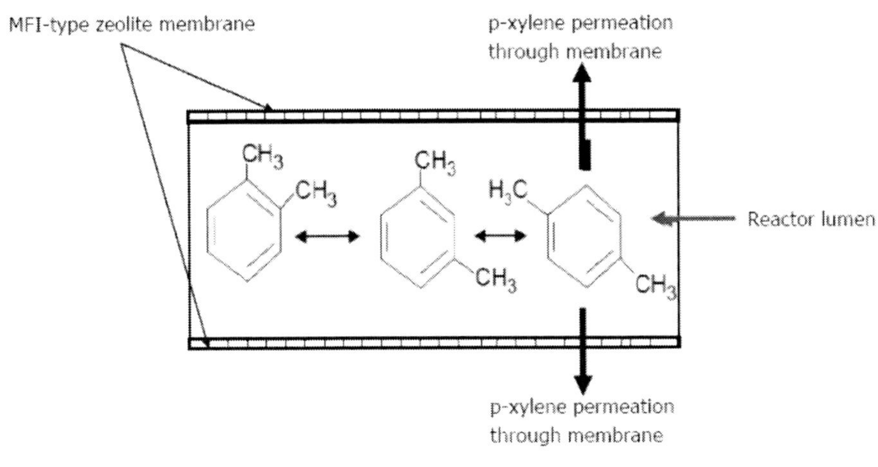

Figure 7. Schematic representation of the selective extraction of *p-Xylene* by a MFI-type zeolite membrane from an isomerization reactor (adapted from ref. [12]).

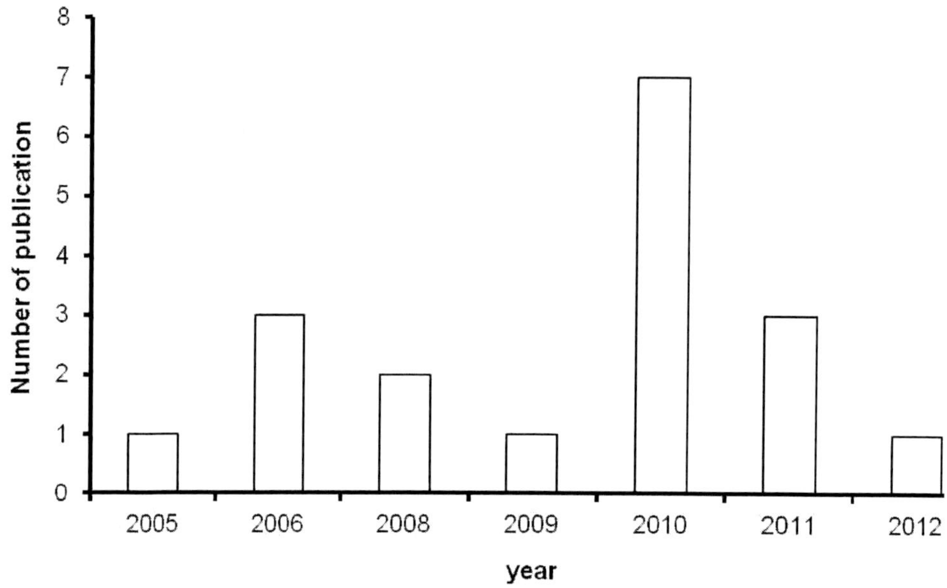

Figure 8. Trends of scientific research on development and applications of e-ZCMRs for xylene synthesis between 2005-2012 (ref. [13])

To date, p-xylene synthesis in e-ZCMRs has been reported with: (i) "thin-film" MFI-type zeolite membranes (TFZ) and, (ii) nanocomposite MFI-type zeolite membranes operated either as a packed-bed membrane reactor or a catalytically active membrane reactor. A TFZ membrane is a supported membrane with thin zeolite layer coated on the top surface of the membrane support. On the contrary, nanocomposite MFI-type zeolite membrane is fabricated using the "so-called" pore-plugging hydrothermal synthesis technique [15-17]. During the pore-plugging hydrothermal synthesis the zeolite crystals are embedded within the pores of the support (see Figure 10 for comparison). In addition, TFZ membranes are prone to cracks and defects at higher temperature applications due to thermal expansion mismatch between the zeolite layer and the support [15-17]. Presence of defects or cracks is undesirable in e-ZCMRs because it leads to loss of para-xylene selectivity, resulting therefore in poor performance of e-ZCMRs. However, this effect is minimized/eradicated in nanocomposite MFI-zeolite membranes because the expansion of the zeolite crystals is limited to the pore spaces that contain the crystals [15-17]. Support usually employed to fabricate MFI-type zeolite membranes include α-alumina [15-17] and some researchers have employed perforated stainless steels as supports [18,19].

Furthermore, an application relying on this combined membrane-catalyst concept for xylene isomerization has been patented by MobilExxon [20]. A general scheme of this process is depicted in Fig. 11. In this process, the fresh feed containing a mixture of xylene isomers is fed into a xylene splitter where the isomers are separated. The exit stream from the splitter (which contains xylenes and EB) enters a xylene recovery unit relying on fractional crystallization and/or molecular sieving to separate p-Xylene from other components. The p-Xylene depleted stream is sent to a membrane reactor to be converted into benzene and/or xylenes, p-Xylene being selectively extracted from the reaction zone using a zeolite membrane.

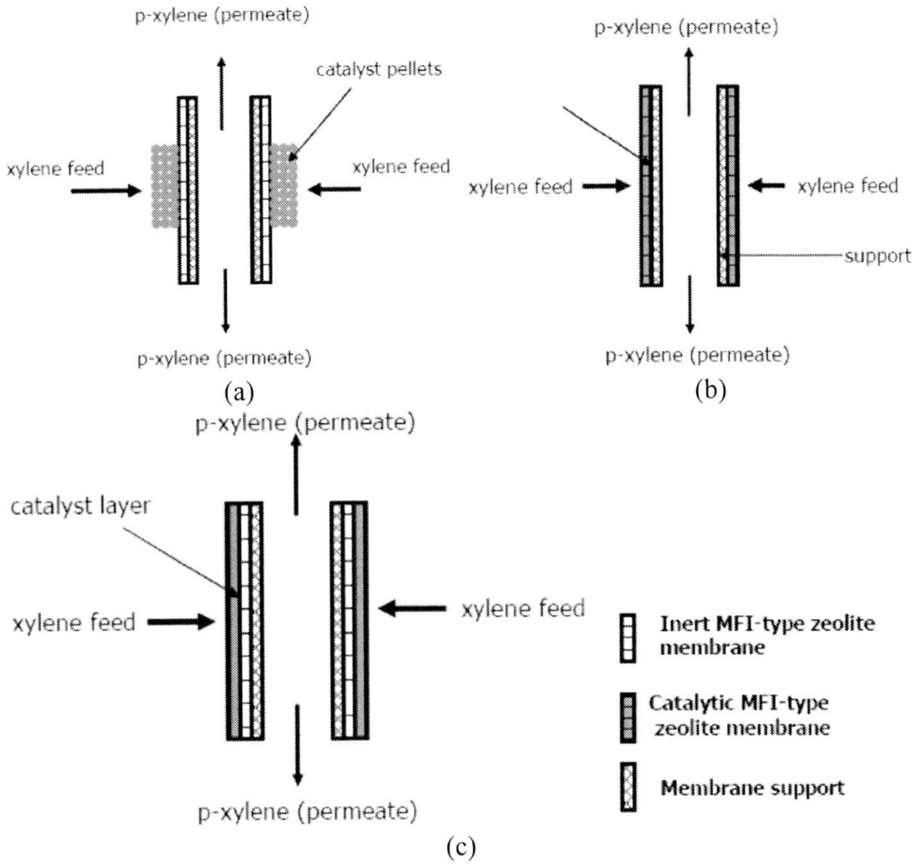

Figure 9. Different extractor-type zeolite CMRs (e-ZCMRs) for xylene isomerization: (a) Inert Zeolite CMR (IZCMR), (b) Active Zeolite CMR (AZCMR) and (c) Bi-functional Zeolite CMR (BZCMR) reactor. Adapted from ref.[14]

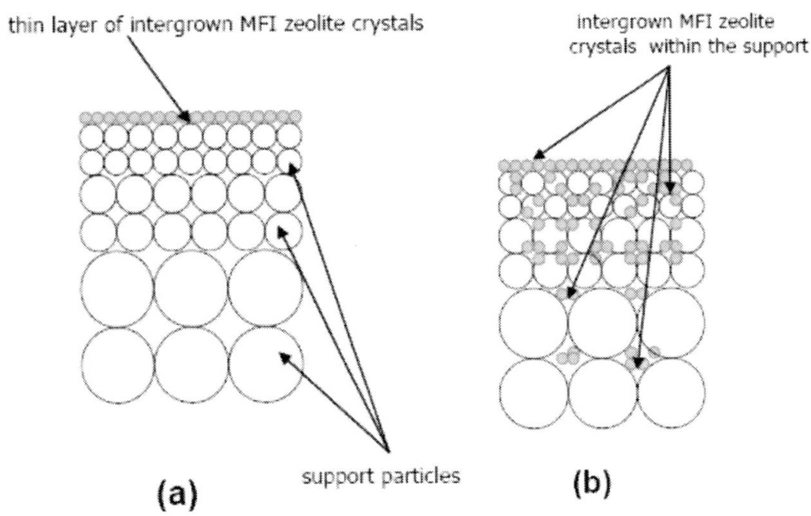

Figure 10. Comparison between TFZ and nanocomposite MFI-type zeolite membranes. TFZ (a) and MFI-type zeolite membrane with nanocomposite architecture (b).

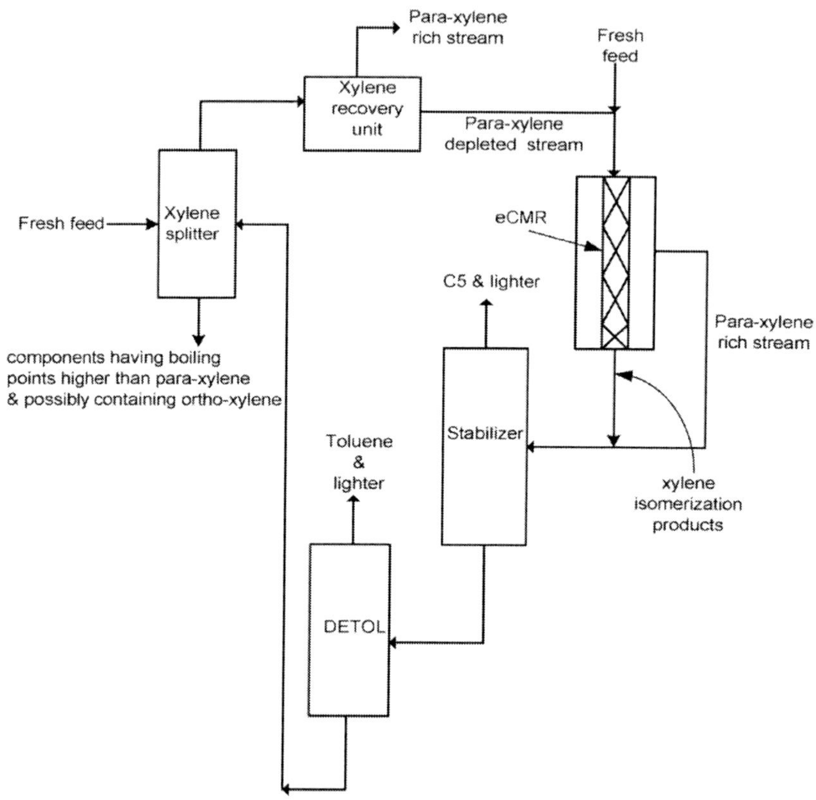

Figure 11. ExxonMobil patented process for p-Xylene production (adapted from refs. [12,20]).

Table 1. Literature survey on xylene isomerization using e- ZCMR based on MFI membranes compared to fixed-bed reactors

Membrane type	T (K)	p-Xylene yield (%)		p-Xylene selectivity (%)		Ref.
		e-CMR	FBR	e-CMR	FBR	
Inert silicalite / α-alumina tube[a]	577	11.2	10.2	65	58	[22]
Inert Ba-ZSM-5/SS[b]	643	25	21	69	52	[18]
Inert silicalite / α-alumina disk[a]	603	21.3	18.5	44.6	35	[21]
Nanocomposite MFI-type zeolite membrane tube	473	26.9	22.9	49.2	41.9	[23,24]
CatalyticH-ZSM-5/SS disk[c]	673	11.8	9.45	-	-	[25]
CatalyticH-ZSM-5/SS disk[a]	673	6.9	5.87	66.7	55.6	[25]
CatalyticH-ZSM-5/SS disk[d]	673	7.3	6.98	30.1	30.1	[25]

Feed composition: [a]m-Xylene feed; [b]ternary mixture of xylene feed; [c]p-Xylene feed; [d]o-Xylene feed.

Regarding the synthesis of p-Xylene via xylene isomerization in extractor-type zeolite catalytic membrane reactor (e-ZCMR), limited studies have appeared in literature (see Table 1). The combination of TFZ membrane with an isomerization catalyst in an e-ZCMR to promote p-Xylene production has been proposed in a few number of previous studies [18,21]. However, TFZ membranes have some shortcomings such as a mismatch between the thermal expansion coefficients of the support and the zeolite material at higher operation temperatures enhancing permeation of undesirable isomers through inter-crystalline defects and contributing to a reduction of membrane selectivity. The occurrence becomes more pronounced if the membrane support is stainless steel because of the higher disparity between its thermal expansion coefficients and that of the MFI phase. Taking into account that xylene separation and isomerization is industrially carried out at temperatures about 673 K, this limitation, in practice, constitutes hindrance for the industrialization of MFI-type zeolite membranes.

Perhaps due to the low quality of the membranes used by these researchers, no appreciable p-Xylene purity could be obtained in the permeate stream. To obtain high-purity p-Xylene, a demanded form of p-Xylene for industrial applications, via conventional processes such as distillation is a difficult and laborious task. Furthermore, the current processes used to achieve this are energy consuming because of the similarity in the physical properties of these isomers [12]. As pointed out in a series of previous studies [15-17,27] and also discussed in Chapter 5 of this book, the shortcoming ascribed to TFZ membranes can be overcome by using nanocomposite MFI-alumina membranes, where the zeolite crystals are embedded within the support pores instead of forming a film. The advantages of nanocomposite architectures compared to zeolite films include defect control and higher mechanical and thermal stability. These membranes have shown promising xylene isomer separation performance with p-xylene/o-xylene separation factor > 400 and p-xylene fluxes in hollow fibre configuration 30% higher than in tubular configurations [22,28,29].

Nanocomposite MFI-Type Zeolite Extractor-Type Catalytic Membrane Reactors for Xylene Production

Regarding the xylene isomerization using MFI-type zeolite membranes with nanocomposite architecture, first open report showed an increase of about 10% p-Xylene yield (when retentate and permeate amounts are combined) over a conventional FBR [22]. Further studies on the use of nanocompoiste MFI-zeolite membranes for para-xylene synthesis were carried out and reported by Daramola et al. [23,24,30]. Therefore, this chapter provides a short overview of previous research efforts on the potential applications of nanocomposite MFI-type zeolite membranes as e-ZCMR for m-Xylene isomerization over Pt-HZSM-5 catalysts and compares their performance with conventional fixed-bed reactors (FBRs).

Interesting and encouraging results have been reported for e-ZCMRs equipped with nanocomposite MFI-type zeolite membranes as separative layers during synthesis of p-Xylene via isomerization of the less used xylene isomers [22-24, 30]. For example in the studies of Daramola et al. [23,24,30], the e-ZCMRs displayed a maximum p-Xylene yield of about 3 % at 473 K when operated in permeate-only mode (only products on the permeate side) and the yield decreased with temperature (see Figure 12 (a)). When the e-ZCMRs were operated at combined mode (addition of the products from the feed side and the permeate side), the p-Xylene yield was about 28%, showing an increase of about 18% over an equivalent fixed-bed reactor operated at similar conditions [23]. Also, the selectivity of the nanocomposite MFI-type zeolite membranes approached 100% towards p-Xylene in permeate-only mode (see Figure 12 (b)) and displayed a maximum value of about 42% at 523 K during the combined mode operation [23,24]. The m-xylene conversion in the e-ZCMR was higher than the conversion in an equivalent FBR operated at similar conditions (see Figure 12 (c)).

On the basis of the high p/o and p/m- xylene separation factors offered by the membrane, xylene compositions of about 95% can be achieved in the permeate stream for both membrane reactor configurations. Additionally, using a defect-free nanocomposite MFI-type zeolite membrane tube with p/o separation factor >400, ultra-pure p-Xylene with xylene composition approaching 100% in the permeate side was obtained even at a lower temperature of 473 K [24]. However, more research efforts are needed in the area of membrane development to increase p-Xylene flux through the membrane while keeping the p-Xylene selectivity at the present values. In addition, loss of reactants (in this case m-xylene) is a major problem in e-ZCMRs. In the work reviewed and discussed in this chapter, about 16-20% of xylene was lost to the formation of by-products like trimethylbenzenes (TMBs) or Toluene (T) or ethylbenzene (EB) via disproportionation of m-Xylene in the retentate side. Interestingly, none of these by-products permeated through the membrane into the permeate side due to the good quality of the nanocomposite MFI-type zeolite membranes applied in these studies. The good quality of these membranes also could be attributed to the nanocomposite architecture of the membranes. This observation is a pointer to the fact that development of e-ZCMR-adaptable high selective catalysts is essential for xylene isomerization in e-ZCMR because the catalyst used in these studies were commercial catalysts originally designed for fixed-bed reactors.

Influence of Operating Conditions on Xylene Production in Extractor-Type Nanocomposite MFI Catalytic Membrane Reactors

The influence of operating variables like reaction temperature, gas hourly space velocity (GHSV = feed volumetric flow per catalyst volume), reaction time and reactor configuration (with respect to catalyst packing) also have been investigated with e-ZCMR equipped with nanocomposite MFI-type zeolite membranes during m-xylene isomerization to p-Xylene [23]. Effect of reactor configuration on the performance of e-ZCMR equipped with nanocomposite MFI-type zeolite membranes was reported [23]. Two different e-ZCMR configurations namely: (i) e-ZCMR-IN (when the catalyst particles are packed inside the lumen of the membrane); (ii) e-ZCMR-OUT (when the catalyst particles are packed within the shell side of the reactor) were considered by the authors (see Figure 13).

When the catalysts were packed inside the membrane lumen an increase of about 18% in the p-Xylene yield at 523 K was obtained in the e-ZCMR over an equivalent conventional FBR due to selective removal of p-Xylene by the membrane (see Figure 14 (a)) [23]. Also for the three configurations considered by the authors (i.e. e-ZCMR-IN, e-ZCMR-OUT and FBR), the p-Xylene yield decreased with temperature in the range 523-673 K (combined mode) while the p-Xylene yield was practically linearly with temperature in e-ZCMR-IN. However, the trend becomes inhibited at lower temperatures for e-ZCMR-OUT and FBR, showing almost a plateau (see Figure 14 (a)). Also, higher m-Xylene conversions and p-Xylene yields were obtained in the e-ZCMR-IN compared to the FBR due to selective p-Xylene extraction from the reaction zone by the membrane (see Figure 14(b)). The m-Xylene conversion reaches a value of about 62% at 523 K in the e-ZCMR-IN while the p-Xylene yield and selectivity attaining values, up to 28% and 46% in combined mode, respectively (see Figure 14(a) & 14 (c)). As depicted in Figure 14 (a), in e-ZCMR-IN molar composition of p-Xylene approached 1.0 at about 600 K and decreased as temperatures increased, but the molar composition of o-Xylene at this temperature was <0.1 (see Figure 15). However at 600 K, molar compositions of p-Xylene and o-Xylene in e-ZCMR-OUT were <0.2. These observations were ascribed to the positive role of the membrane in p-Xylene extraction in the e-ZCMR-IN. In the e-ZCMR-OUT, the trend of the p-Xylene yield with the temperature suggests a reduced effect of the membrane on the catalytic performance due to the diffusion limitations in the membrane support, as permeation proceeds from the outer surface to the inner top layer. Similar conclusions was drawn by Zhang et al. in their studies involving the application of TFZ membrane with membrane lumen packed with H-ZSM-5 commercial catalysts for synthesis of p-Xylene via m-xylene isomerization. [21]. The higher p-Xylene yield at lower temperatures (i.e. 523 K) was attributed to a higher concentration between the membrane and catalyst activities, providing a higher efficiency of the membrane reactor and consequently a higher differential in terms of p-Xylene yield compared to a FBR.. Moreover, lower temperatures are also beneficial due to the reduction of formation of by-products (i.e. toluene, trimethylbenzene and ethybenzene).

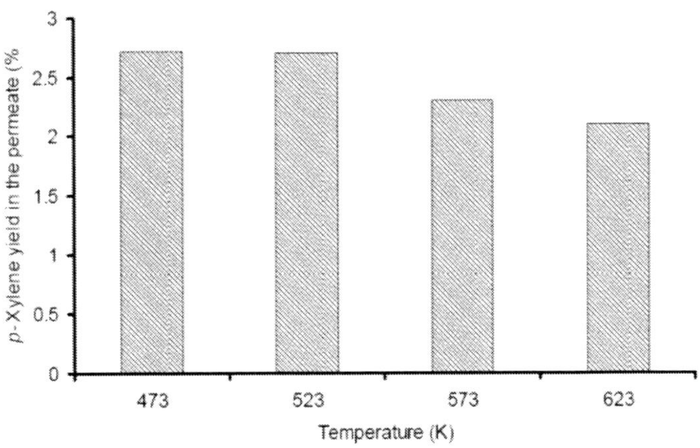

(a)

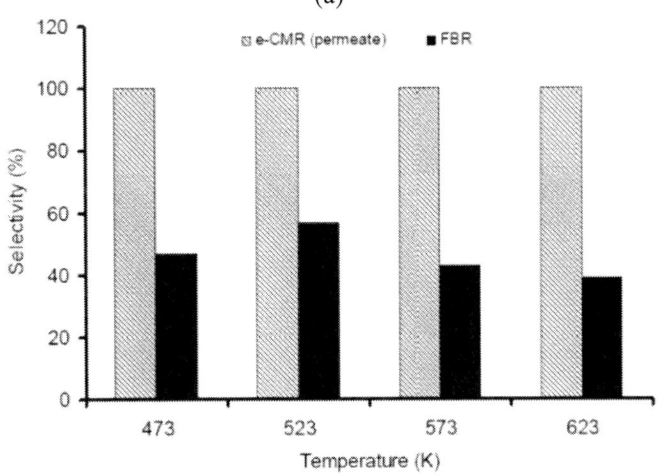

(b)

Figure 12. (Continued)

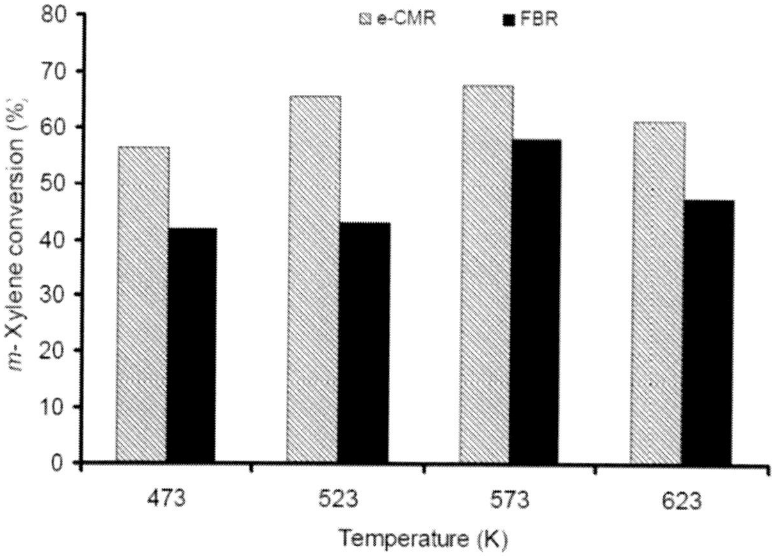

(c)

Figure 12. Performance of e-ZCMR equipped with nanocomposite MFI-type zeolite membranes compared with an equivalent FBR as a function of temperature during meta-xylene isomerization to p-xylene over Pt-HZSM-5 catalyst (a) p-xylene yield in permeate only mode (b) p-xylene selectivity (para-selectivity); (c) meta-xylene conversion (adapted from refs. [23,24]).

Regarding the effect of gas-hourly-space-velocity (GHSV) and reaction time on the performance of e-ZCMR equipped with nanocomposite MFI-type zeolite membranes during isomerization of m-xylene into p-xylene, no remarkable influence on either the p-Xylene yield or the selectivity was reported [23]. This observation is consistent with report of Zhang et al. when TFZ membrane was used as e-ZCMR to isomerize m-Xylene into p-Xylene [21]. The behaviors of e-ZCMR with changes in reaction time and GHSV can be explained in terms of a reduced influence of p-Xylene production through an intermolecular transalkylation mechanism responsible for toluene and trimethylbenzene formation, the diphenylmethane intermediates of this reaction acting as promoters of coke formation on strong acid sites [31].

Catalytically Active Extractor-Type Membrane Reactors for Xylene Synthesis

In the case of the application catalytically active MFI-type zeolite membranes for xylene synthesis, a few reports has appeared in literature [25,26]. Haag and co-workers [25] has used catalytically active MFI-type zeolite membrane consisting of a HZSM-5 layer grown on top of a porous stainless steel (SS) disk during isomerization of m-Xylene to p-Xylene. The authors reported an increase of 15% in m-Xylene conversion at 673 K compared to an equivalent FBR, while p-Xylene selectivity was increased by 10%. However at higher temperatures, no further improvement in the conversion or p-Xylene selectivity was achieved by the authors, probably due to presence of more intercrystalline defects in their membrane, resulting to poor separation quality of the membrane. Recently, a report by Zhang et al. reported the application of catalytically active MFI-type zeolite membranes to xylene isomerization. The membranes were synthesized by performing H+ ion-exchange treatment on MFI/ α-Al_2O_3 membrane [26]. The authors claimed to obtain ~92% p-Xylene selectivity during xylene isomerization. However, the m-Xylene conversions of these membranes were very low due to limited number and non-uniform distribution of the catalytically active sites on the membranes.

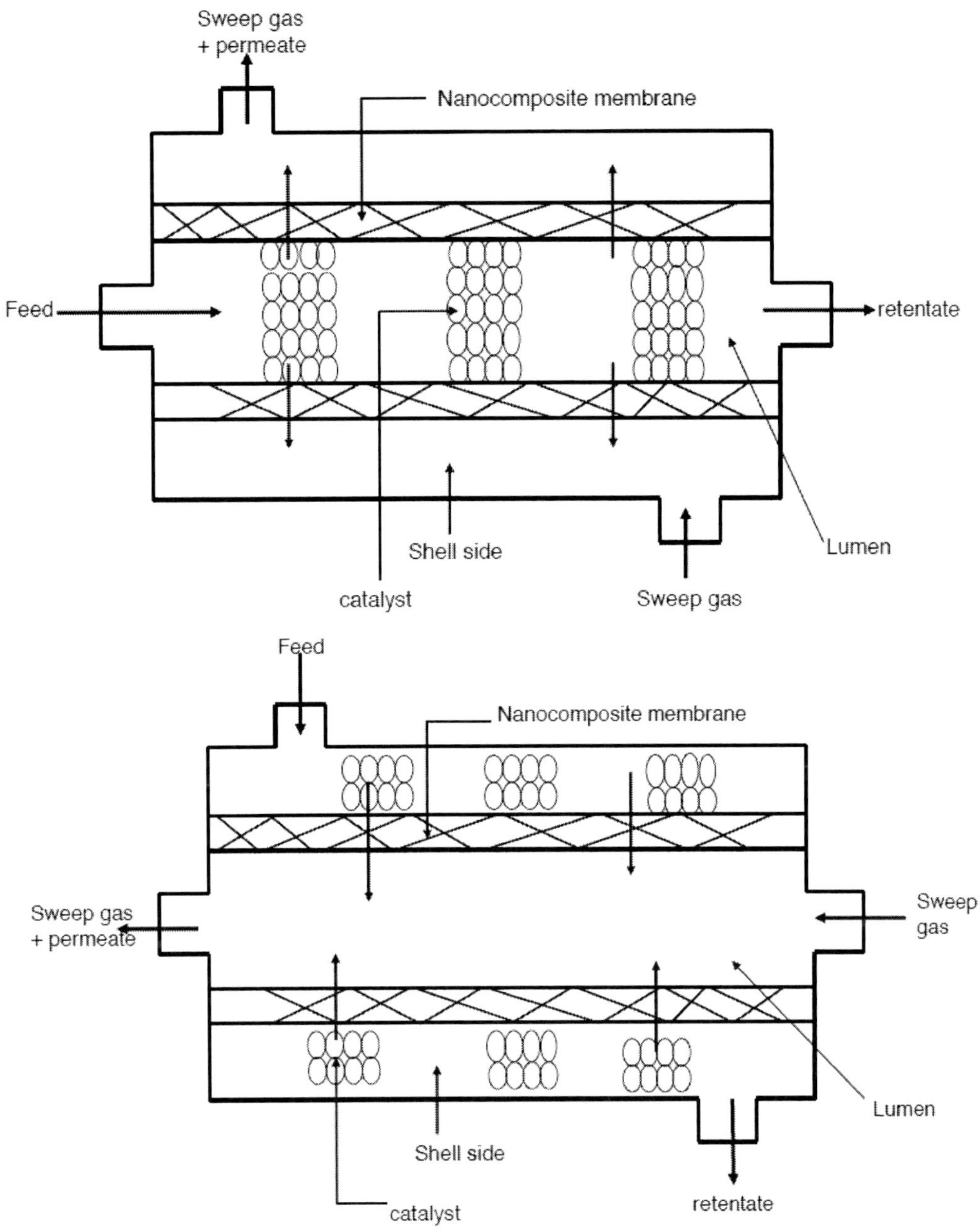

Figure 13. Types of reactor configurations: e-ZCMR-IN (top), e-ZCMR-OUT (bottom).

Catalytically active MFI-type membranes might be a good option for membrane-based synthesis of xylenes because they could be about ten (10) times more active than when catalyst pellets are used in the fixed-bed reactor, but getting uniformly dispersed catalyst active site within the membranes could be a major problem [32,33]. In catalytically active membrane, the membrane thickness and porous texture, as well as the quantity and location of the catalyst in the membrane are essential and they should adapt to the reaction kinetics [32,33]. Therefore concerted research efforts are expected to be geared towards the synthesis and application of catalytically active membranes in the near future.

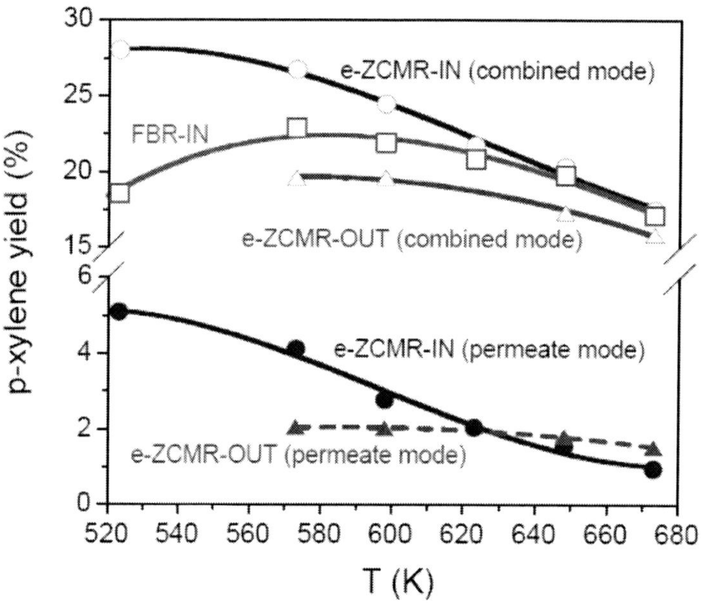

(a)

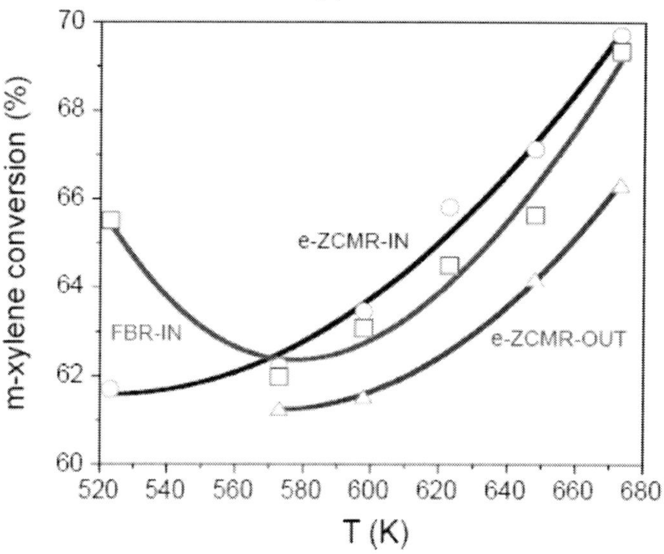

(b)

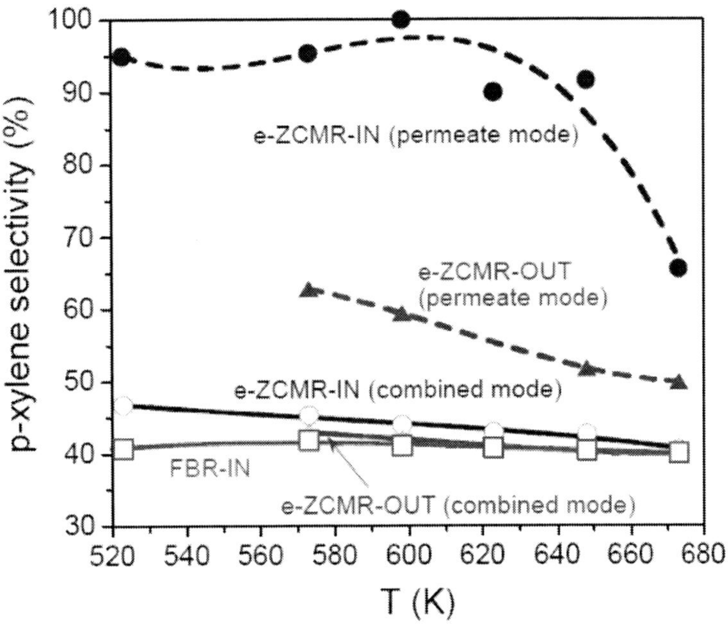

(c)

Figure14. Influence of operating conditions on the performance of e-ZCMR equipped with nanocomposite MFI-type zeolite membranes during isomerisation of m-Xylene to p-xylene. (a) p-xylene yield; (b) m-xylene conversion (c) p-xylene selectivity. (Obtained with permission from ref. [23]).

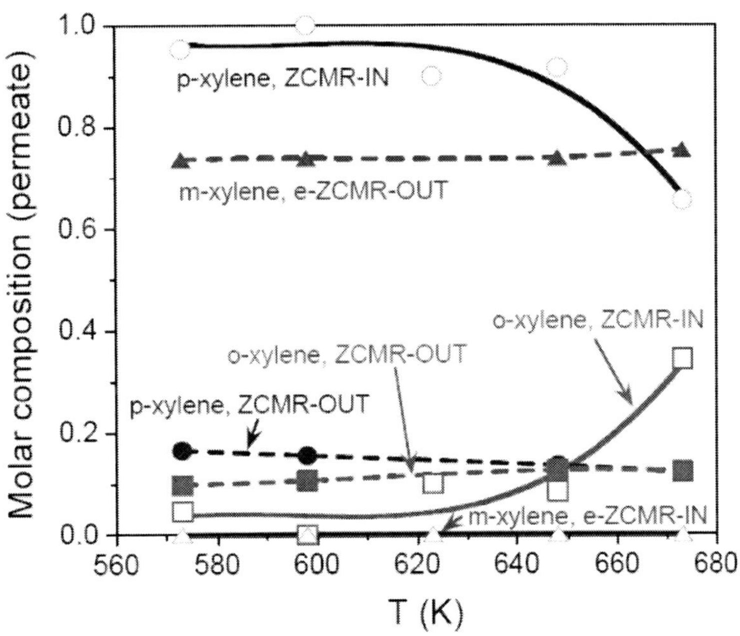

Figure 15. p-Xylene, m-Xylene and o-Xylene molar composition in permeate in ZCMR-IN and ZCMR-OUT configurations as a function of temperature. (Obtained with permission from ref.[23])

CONCLUSION

The results reviewed and described in this chapter showcase excellent performance of e-ZCMRs for m-Xylene isomerization to p-Xylene over the conventional FBRs. With the use of a defect-free nanocomposite MFI-type membrane tube displaying p/o separation factor >400, possibility of obtaining ultra-pure p-Xylene with xylene composition approaching 100% in the permeate side at one stage, even at a lower temperature of 473 K, has been demonstrated [24]. However, fluxes of these membranes are still too low to attract industrial applications. Also, various experimental tests reported in literature were conducted at operating conditions (expect reaction temperatures) relatively low compared to operating conditions employed in Industry. Therefore more research efforts are needed in terms of membrane development and optimization to increase p-Xylene flux through the membrane while keeping the p-Xylene selectivity at present values. Also experimental evaluation of these membranes at industrial operating conditions is essential. Additionally, it has been pointed out that e-ZCMR is associated with loss of xylenes during xylene synthesis. For instance, Daramola et al. observed loss in xylene during m-xylene isomerization in e-ZCMR equipped with nanocomposite MFI-type zeolite membranes due to formation of by-products such as trimethylbenzenes (TMBs) or Toluene (T) or ethylbenzene (EB) via disproportionation of m-Xylene in the retentate side but these membranes still maintained their selectivities to p-Xylene [23,24]. However, this observation is a pointer to the fact that development of e-ZCMR-adaptable high selective catalysts is essential for xylene isomerization in e-ZCMR since various catalysts used in the reports described in literature were commercial catalysts originally designed for fixed-bed reactor.

With the excellent performance of e-ZCMRs during xylene synthesis in comparison to the conventional FBRs, the possibility of cutting down operational cost via a reduction in energy consumption during xylene synthesis could be feasible with the application of e-CMR because no additional units are needed to get ultra-pure p-Xylene. However, "defect-free" MFI-type zeolite membranes with reasonable p-Xylene fluxes are necessary to make this technology attractive and competitive with existing technologies. In addition, higher surface-to-volume ratio of the nanocomposite MFI-type zeolite hollow fibre membranes can be exploited toward enhancing p-Xylene fluxes [28]. Also, these hollow fibre membranes could be optimized into fibre modules reducing therefore the membrane module layouts. Therefore, excellent separation potential of nanocomposite MFI-type zeolite membranes (tubes and hollow fibres) as e-IZCMRs could catalyse the development of energy-efficient membrane-based process for one-stage production of high purity p-Xylene. Furthermore, with the new research line on the fabrication of mixed matrix membranes with zeolite crystals as fillers, great milestones in the development and application of e-ZCMRs will be experienced in the future.

ACKNOWLEDGMENT

The author acknowledges Elsevier publisher BV, The Netherlands for the permission to use some figures from its published work in this chapter.

REFERENCES

[1] Chirico, RD; Steele, WV. *J. Chem. Eng. Data*, 42(4), (1997), 784-790.

[2] International Energy outlook 2009 report, Energy Information Administration, Office of integrated Analysis and Forecasting, U.S. Department of Energy, Washington, DC 20585, *DOE/EIA-0484*, 2009; 284.

[3] Siirola, JJ. *AIChE Symposium Series*, 91(304), (1995), 222-233.

[4] Daramola, MO; Aransiola, EF; Ojumu, VT. *Materials*, 5 (2012), 2101-2136.

[5] Koros, WJ; Ma, YH; Shimidzu, T. *Pure Appl Chem*. 68(7), (1996), 479-1489.

[6] Cao, P; Tremblay, AY; Dubé, MA; Morse, K. *Industrial & Engineering Chemistry Research*, 46(1), (2007), 52-58.

[7] Scott, FH. *Elements of Chemical Reaction Engineering*, 4rd Ed. Prentice-Hall: Upper Saddle River, NJ, 2005.

[8] Miachon, S; Dalmon, J-A. *Topics in Catalysis*, 29(1-2), (2004), 59-65.

[9] Hsieh, HP. *Inorganic Membranes for separation and reaction*, Elsevier, Amsterdam, 1996

[10] Reif, M; Dittmeyer, R. *Catal. Today*, 82, (2003), 3-14.

[11] Armor, JN. *J. Membr. Sci*. 147(2), (1998), 217-233.

[12] Daramola, MO; Burger, AJ, Pera-Titus, M; Giroir-Fendler, A; Miachon, S; Dalmon, JA; Lorenzen, L. *Asia-Pacific Journal of Chemical Engineering*, 5(6), (2010), 815-837

[13] www.scopus.com.

[14] Mabande, GTP; Noack, M; Avhale, A; Kolsch, P; Georgi, G; Schwieger, W; Caro, J. *Micropor. Mesopor. Mater*. 98(1-3), (2007), 55-61.

[15] Miachon, S; Landrivon, E; Aouine, M; Sun, Y; Kumakiri, I; Li, Y; Pachtova Prokopova, O; Guilhaume, N; Giroir-Fendler, A; Mozzanega, H; Dalmon, J-A. *J. Membr. Sci*. 281, (2006), 228-238.

[16] Miachon, S; Ciavarella, P; van Dyk, L; Kumakiri, I; Fiaty, K; Schuurman, Y; Dalmon, J-A. *J. Membr. Sci*. 298 (2007), 71-79.

[17] Li, Y; Pera-Titus, M; Xiong, G; Yang, W; Landrivon, E; Miachon, S; Dalmon, J-A. *J. Membr. Sci*. 325, (2008), 973-981.

[18] Tarditi, AM; Horowitz, GI; Lombardo, EA. *Catal Lett*. 123, (2008), 7-15.

[19] Tarditi, AM; Lombardo, EA. *Sep. Purif. Technol*. 61, (2008), 136-147.

[20] Ferraro, JM; Osman, RM; Ou, D-Y. US Patent 6376733 B1, *ExxonMobil Chemicals*, Houston, Texas, 2002.

[21] Zhang, C; Hong, Z; Gu, X; Zhong, Z; Jin, W; Xu, N. *Ind. Eng. Chem. Res*. 48, (2009), 4293-4299.

[22] Van Dyk, L; Lorenzen, L; Miachon, S; Dalmon, J-A. *Catal. Today*, 104, (2005), 274-280.

[23] Daramola, MO; Deng, Z; Pera-Titus, M; Giroir-Fendler, A; Burger, AJ; Lorenzen, L; Guo, Y. *Catalysis Today*, 156, (2010), 261-267.

[24] Daramola, MO; Burger, AJ; Miachon, S; Giroir-Fendler, A; Lorenzen, L. *Appl. Catal. A: General*, 386, (2010), 109-115.

[25] Haag, S; Hanebuth, SM; Mabande, GTP; Avhale, A; Schwieger, W; Dittmeyer, R. *Micropor. Mesopor. Mater*. 96, (2006), 168-176.

[26] Zhang, C; Hong, Z; Chen, J; Gu, X; Jin, W; Xu, N. *J. Membr. Sci.* 389, (2012), 451-458.

[27] Giroir-Fendler, A ; Peureux, J; Mozzanega, H; Dalmon, J-A. *Stud. Surf. Sci. Catal.* 101, (1996), 127-136.

[28] Daramola, MO; Burger, AJ; Pera-Titus, M; Giroir-Fendler, A; Lorenzen, L; Miachon, S; Dalmon, J-A. *J. Membr. Sci.*, 337, (2009), 106-112.

[29] Daramola, MO; Burger, AJ; Pera-Titus, M; Giroir-Fendler, A; Lorenzen, L; Dalmon, J-A. *Sep. Sci. Technol.*, 45(1), (2010), 21-27.

[30] Daramola, MO; Burger, AJ; Giroir-Fendler, A. *Chemical Engineering J.* 171, (2011), 618-627.

[31] Guisnet, M; Gnep, NS; Morin, S. *Micropor. Mesopor. Mater.* 35-36 (2000), 47-59.

[32] Harold, MP; Lee, C; Burggraaf, AJ; Keizer, K; Zaspalis, VT; De Lange, RSA. *MRS Bull.* 19(4), (1994), 34-39.

[33] Zaspalis, VT; Van Praag, W; Keizer, K; Van Ommen, JG; Ross, JRH; Burggraaf, AJ. *Appl. Catal.* 74(2), (1991), 205-222.

In: Xylenes
Editor: Michael Olawale Daramola
ISBN: 978-1-62808-342-2

Chapter 7

Synthesis of Nanocomposite Boron-Substituted MFI-Type Zeolite Membranes and Application to Xylene Isomers Separation

***Z. Deng*[1] *and M. O. Daramola*[2]**
[1]Chengdu Institute of Organic Chemistry, Chinese Academy of Science, Chengdu, P. R. China
[2]School of Chemical and Metallurgical Engineering, Faculty of Engineering & the Built Environment, University of the Witwatersrand, Johannesburg, South Africa

Abstract

In this chapter, synthesis of Boron substituted nanocomposite MFI-type zeolite membranes (B-MFI) and application of the membranes to xylene isomers separation is presented. The membranes were prepared through "pore-plugging" hydrothermal synthesis technique on porous supports. In terms of separation performance, B-MFI membranes have displayed better performance when compared with un-substituted nanocomposite MFI zeolite membranes.

Keywords: Membranes, boron, xylenes, separation, MFI

Introduction

The isomorphous substitution of Al and/or Si by other metal ions such as B, Ga, Fe, V, Ti, Ge and Ce is generally used for tailoring the adsorption and catalytic properties of MFI zeolites [1, 2]. In addition, substitution of Al or Si could result in modification of the chemical properties (e.g., hydrophobicity and Brönsted acidity) and channel size of MFI zeolites due to the change of T-O-T bond angles and T-O bond lengths (T is Si, Al or substituted element atom). The Brönsted acidity strength for substituted zeolites has been reported to increase in the order: silicalite-1 < B-MFI < Fe-MFI < Ga-MFI < Al-MFI [1]. In the same vein, separation performance of un-substituted MFI zeolite membrane could be enhanced through isomorphous substitution. Various substituted MFI-type zeolite membranes like Ti-MFI [3-4], V-MFI [5-7], Fe-MFI [8, 9], Ce-MFI [10] and Ge-MFI [11, 12] have been synthesized and applied as membranes in separation applications and in reactive separation using catalytic membrane reactors.

In previous chapters, synthesis and application of un-substituted MFI-type zeolite membranes in xylene isomers separation have been enumerated. However in this chapter, the focus is on boron substituted MFI-type zeolite membranes, and their potential application in xylene isomers separation. Compared with Al and Si, the ionic radius of B is smaller, expectedly B-MFI has smaller channels, provides more rigid MFI framework, as well as weaker Brönsted acid sites, which further affect the separation performance of the substituted zeolite membranes. Incorporation of number of boron atoms into the MFI framework can be controlled by using different templates during the hydrothermal synthesis. About 2,4 and 10~14 boron atoms per unit cell of MFI framework have been reported by Pergo et *al* using tetrapropylammonium (TPA), tetrapentylammonium (TPeA) and ethylenediamine(EN) cations as templates [13].

SYNTHESIS AND CHARACTERIZATION OF BORON-EXCHANGED MFI-TYPE ZEOLITE MEMBRANES

Incorporation of Boron into the MFI framework to replace Al or Si in MFI zeolite is either by post-synthesis impregnation of MFI zeolite with an aqueous solution of H_3BO_3 [14] or treatment of the zeolite with BCl_3 [15]. Alternatively, the B-MFI membrane can be prepared by hydrothermal synthesis using a precursor solution containing silica source, H_3BO_3 as boron source and TPAOH as template. Supports that can be used to prepare supported zeolite membranes include α-aluminum or porous stainless steel either in tubular, monolith or hollow fibre configurations [16-19]. Usually after the synthesis, the synthesized membranes are washed until pH of 7 and dried at 100°C in an oven overnight. After drying, the as-prepared templated MFI membranes are normally calcined (at about 500°C for 5 hours with a temperature ramp rate 1°C/min) to remove the organic template occluded inside the pores of the zeolite.

In the case of B-MFI membranes, the qualitative analysis to probe the presence of boron (B) in the membrane was performed on solid-state magic angle spinning nuclear magnetic resonance spectroscopy (MAS-NMR). In the case of the ^{11}B MAS-NMR spectra, the absence of peak centers at -1 ppm suggests the absence of boron occupying trigonal framework positions, and the absence of peak centers at 15 ppm suggests the absence of extra-framework boron ascribed to amorphous boron oxides grown in the MFI channels. Also, the peak centers at -4 ppm is the characteristic peak of tetrahedral Si-O-B, as depicted in Figure 1 [16, 20, 21]. The quantitative analysis of the Si/B ratio was obtained from inductively coupled plasma atomic emission spectroscopy (ICP-AES). Phase purity of the synthesized B-MFI membranes according to the characteristic peaks was checked using X-ray diffraction(XRD) and Fourier transform infrared (FTIR), also compared with MFI membranes, as depicted in Figure 2.

Also, the morphology of the synthesized B-MFI membrane was checked with scanning electron microscope (SEM).

In the case of thin film zeolite membranes (TFZ), membrane thickness of the membranes is obtainable directly from the acquired SEM images of the membranes because the thin film zeolite crystal layer is only present on the surface of the support. In some cases, especially for hollow fibre and tubular membranes, the membranes are carefully broken and the broken parts prepared for SEM analysis. However for nonacomposite zeolite membrane, it is impossible to measure membrane thickness directly from the acquired SEM images because the zeolite crystals are embedded within the pores of support. The SEM images of hollow fibre support and nanocomposite B-MFI membrane are shown in Figure 3.

Therefore, membrane thickness of nanocomposite B-MFI membrane was obtained indirectly using single gas permeation technique and Maxwell-Stefan (MS) model (see Equation 1) [22, 23]. The advantage of this method is that the membrane is intact. Typical curves of single gas permeation tests (N_2 and CO_2) as a function of temperature (298K~673K) for nanocomposite B-MFI-type zeolite membranes are depicted in Figure 4. To obtain the membrane thickness using the aforementioned technique, data acquired from the single gas permeation tests is fitted with MS model (Equation 1).

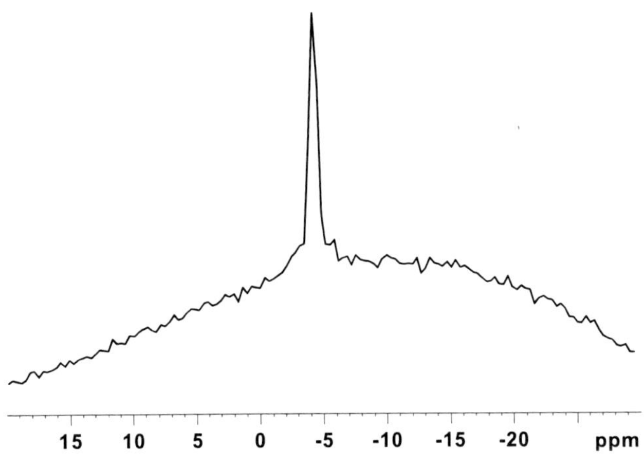

Figure 1. ^{11}B MAS-NMR spectra of the B-MFI crystals (adapted from ref.[16]).

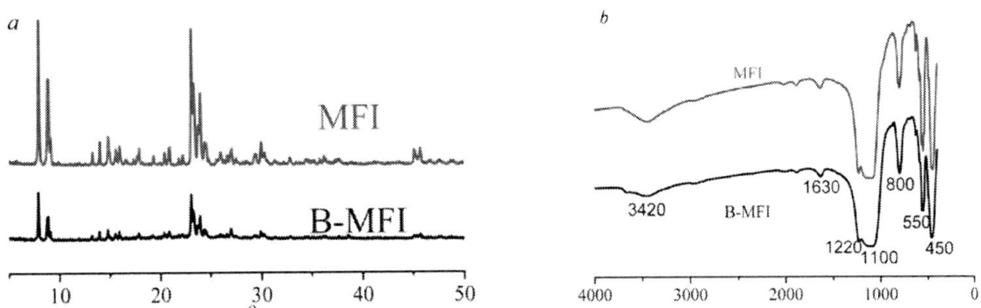

Figure 2. XRD patterns (a) and FT-IR spectra (b) of MFI and B-MFI membranes (adapted from ref.[16]).

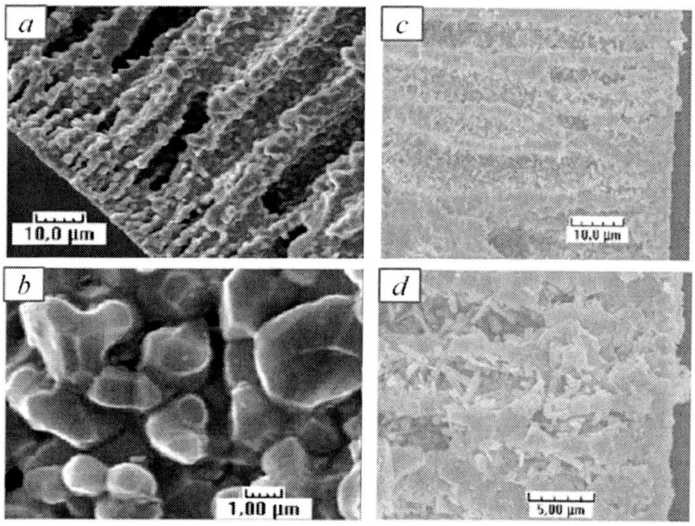

Figure 3. SEM micrographs of cross section of hollow fibre (left) and nanocomposite B-MFI hollow fibre membrane(right) with increasing magnification from top to bottom. (adapted from ref. [16]).

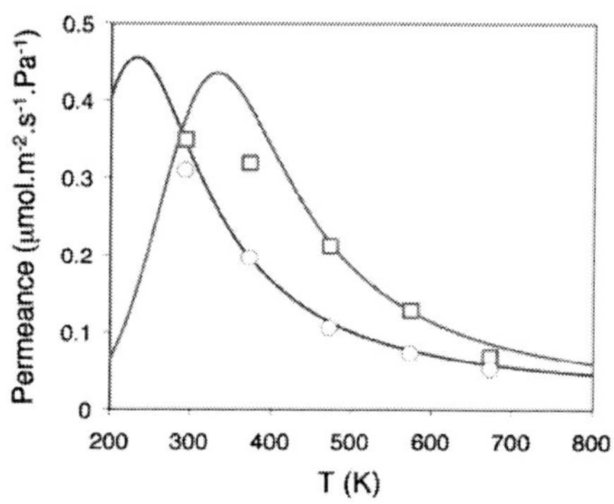

Figure 4. Single gas N_2(○) and CO_2(□) permeation on B-MFI nanocomposite membrane under different temperature, (adapted from ref. [16,24]).

$$N = \frac{c_{sat}\rho\varepsilon D_o^{\infty}}{\tau\ell}\ln\left[\frac{1+\frac{P_R}{P^o}\exp\left(\frac{\Delta S^o_{ads}-R}{R}-\frac{\Delta H^o_{ads}}{RT}\right)}{1+\frac{P_P}{P^o}\exp\left(\frac{\Delta S^o_{ads}-R}{R}-\frac{\Delta H^o_{ads}}{RT}\right)}\right]\exp\left[-\frac{E_D}{RT}\right] \qquad (1)$$

And the gas permeance, N, is obtained from equation (2):

$$N = \frac{Q_g}{A_m \times \Delta P} \qquad (2)$$

Where Q_g, is the gas permeation rate through the membrane in mol.s-1; A_m, the membrane area in m^2; and ΔP, the transmembrane pressure in Pa; N, the gas permeance in

$mol.s^{-1}.m^{-2}.Pa^{-1}$; R, the ideal gas constant; T, the temperature in K; C_{sat}, saturation concentration of the gas in MFI crystals, ρ_{MFI}, density of MFI; ε, the porosity of the nanocomposite MFI / alumina structure; D_0^{∞}, the Maxwell-Stefan diffusivity at zero coverage, τ, the membrane tortuosity, ℓ, the equivalent MFI thickness, P_R, the retentate pressure; P_P, the permeate pressure; P^o, the atmospheric pressure; ΔS^o_{ads}, the standard adsorption entropy; ΔH^o_{ads}, the standard adsorption enthalpy; and E_D, the diffusion activation energy.

Application to Xylene Isomers Separation

Depending on the phase state of feed, separation of xylene isomers is done by pervaporation (PV) or vapor permeation (VP). Separation by PV is considered when the feed is in liquid state and VP when the feed is in vapour phase. During separation by PV, concentration polarization due to high concentration of xylene is a major problem [25-28]. Regarding the application of B-MFI-type zeolite membranes to separation of xylene isomers, Lin et *al*. investigated effect of B/Si ratios (10, 50 and 100) on the separation performance of thin film B-MFI membranes during pervaporative separation of xylene isomers [29]. The authors reported that p-xylene/o-xylene (p/o) separation factor increased at increasing B/Si ratio, indicating more boron in the framework of zeolite. Increasse in p/o separation factor was attributed to a decrease in the volume of unit cell and also the enhancement in the rigidity of the MFI framework. However at increasing xylene feed concentration, loss in membrane selectivity was observed as a result of higher p-xylene concentration in the feed (80 wt.% p-xylene, 20 wt.% o-xylene). The authors reported highest p/o separation factor of 55 at 5 wt.% p-xylene in the feed mixture. This observation implies that distortion of MFI framework occurs at higher p-xylene loadings during PV, allowing passage of o-xylene and thereby making the membrane to lose its separation performance very fast. Over a period of 48 hours, the separation performance of the B-MFI membranes became unstable and the p/o separation factor decreased dramatically (a reduction of about an order of magnitude).

In order to avoid the distortion of zeolite framework that is experienced during PV at higher p-xylene loading, separation of xylene isomers at low pressure by VP has been extensively studied [30-33]. In the same vein, nanocomposite B-MFI-type hollow fibre membranes have been synthesized and tested during xylene isomers separation to evaluate their separation performance and behaviors during VP of xylene isomers [16]. Nanocomposite B-MFI hollow fibre membrane has some advantages over other membrane structures reported in literatures, the hollow fibre support has higher separation surface area/volume ratio compared to tubular suppport which allow for more compact and cost-effective units. In addition, enhanced permeation flux, attributed to the thinner wall thickness of the support, is provided by the configuration [32]. Also, the nanocomposite structure formed via the pore-plugging technique, which the zeolite crystals are embedded within the pores of supports, minimizes formation of cracks that occurs in TFZ at higher temperatures. Formation of these cracks in TFZ is attributed to thermal expansion mismatch between zeolite crystals and support. Also effects of abrasion during membranes handling and module assembling could be eliminated [22,23].

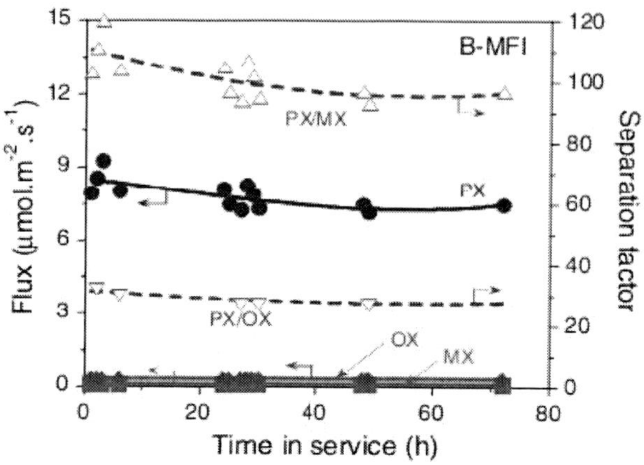

Figure 5. Continually operated at 473 K and for a feed partial pressure PX 0.7 kPa, MX 0.8 kPa, OX 0.8 kPa on nanocomposite B-MFI hollow fibre membrane, (adapted from ref. [16]).

When the nanocomposite B-MFI hollow fibre membranes were employed in xylene isomers separation via vapor separation, the separation performance was improved in comparison to the non-substituted nanocomposite MFI hollow fibre membrane [16]. First, Firstly, p/o and p-xylene/meta-xylene (p/m) separation factor increased as the permeance of other competing isomers decreased dramatically. Also, the B-MFI hollow fibre membrane still maintained its performance at a slightly higher feed vapor pressure of xylene (4.2 kPa), a better separation performance over a non-substituted nanocomposite MFI hollow fibre membrane and TFZ [24, 32, 34-36]. Third, B-MFI hollow fibre membrane is still stable after continuous operation for 72 hours, as showing in Figure 5.

CONCLUSION

Overall, synthesis and characterization of B-MFI membranes have been discussed in this chapter. Also, application of the synthesized membranes to separation of xylene isomers has been showcased. Nanocomposite B-MFI membranes synthesized via pore-plugging hydrothermal synthesis displayed better performance when compared with thin film B-MFI membranes prepared by traditional hydrothermal synthesis technique. While separation performance of MFI-type zeolite membranes during xylene isomers separation could be enhanced by incorporating B into MFI framework, distortion of MFI framework could not be completely eliminated via isomorphous substitution. In addition, B-MFI zeolite membranes have shown enhanced separation performance during separation of other mixtures like hexane/2,2-dimethylbutane [37], C_8 aromatic mixtures [38], organic alcohol/water [39,40], CO_2/N_2 [41] and so on. However, further concerted efforts are required in the area of membrane development and optimization to produce high-flux B-MFI membranes that could be applicable at very high loading of xylene isomers, preferably using PV. Availability of such membranes could fast-track commercial development of membrane-based process for xylene synthesis.

REFERENCES

[1] Szostak, R. Molecular Sieves: Principles of Synthesis and Identification, *Van Nostrand Reinhold*, New York, 1989.

[2] Perego, G; Millini, R; Bellussi, G. in: H.G. Karge, J. Weitkamp, (Eds.), *Molecular Sieves – Science and Technology*, vol. 1 Synthesis, Springer-Verlag, Heidelberg, 1998.

[3] Motuzas, J; Mikutaviciute, R; Gerardin, E; Julbe, A. Controlled growth of thin and uniform TS-1 membranes by MW-assisted heating, *Micropor. Mesopor. Mater.*, 128, (2010), 136-143.

[4] Au, LTY; Chau, JLH; Tellez, C; Yeung, KL. Preparation of supported Sil-1, TS-1 and VS-1 membranes effects of Ti and V metal ions on the membrane synthesis and permeation properties, *J. Membr. Sci.*, 183, (2001), 269-291.

[5] Julbe, A; Farrusseng, D; Guizard, C; Jalibert, J; Mirodatos, C. Characteristics and performance in the oxidative dehydrogenation of propane of MFI and V-MFI zeolite membranes, *Catal. Today*, 56, (2000), 199-209.

[6] Julbe, A; Farrusseng, D; de Ménorval, LC; Guizard, C. Characterization of MFI/α-Al_2O_3 and V-MFI/α-Al_2O_3 composite membranes by ^{129}Xe NMR, *Sep. Purif. Technol.*, 32, (2003), 165-173.

[7] Tavolaro, A. VS-1 composite membrane: preparation and characterization, *Desalination*, 147, (2002), 333-338.

[8] Zhang, X; Li, Y; Wang, J; Tong, H; Liu, C. Synthesis and characterization of Fe-MFI zeolite membrane on a porous α-Al_2O_3 tube, *Sep. Purif. Technol.*, 25, (2001), 269-274.

[9] Zhang, X; Wang, J; Liu, H; Liu, C; Yeung, KL. Factors affecting the synthesis of heteroatom zeolite Fe-ZSM-5 membrane, *Sep. Purif. Technol.*, 32, (2003), 151-158.

[10] Cheng, Y; Li, J-S; Wang, L-J; Sun, X-Y; Liu, X-D. Synthesis and characterization of Ce-ZSM-5 zeolite membranes, *Sep. Purif. Technol.*, 51, (2006), 210-218.

[11] Tuan, VA; Noble, RD; Falconer, JL. Isomorphous substitution of Al, Fe, B and Ge into MFI-zeolite membranes, *Micropor. Mesopor. Mater.*, 41, (2000), 269-280.

[12] Li, S; Tuan, VA; Falconer, JL; Noble, RD. Properties and separation performance of Ge-ZSM-5 membranes, *Micropor. Mesopor. Mater.*, 58, (2003), 137-154.

[13] Perego, G; Bellussi, G; Millini, R; Alberti, A; Zanardi, S. B-containing molecular sieves crystallized in the presence of ethylenediamine. Part II: crystal structure of as-synthesized B-MFI, *Micropor. Mesopor. Mater.*, 58, (2003), 213-223.

[14] Kaeding, WW; Chu, C; Young, WB; Weinstein, B; Butter, SA. Selective alkylation of toluene with methanol to produce para-Xylene, *J. Catal.*, 67, (1981), 159-174.

[15] Dessau, RM; Schmitt, KD; Imelik, B; Naccache, C; Coudurier, G; Ben Taarit, Y; Védrine, JC (Eds.). *Catalysis by Acids and Bases*, Elsevier, Amsterdam, 1985.

[16] Deng, Z; Nicolas, C-H; Guo, Y; Giroir-Fendler, A; Pera-Titus, M. Synthesis and characterization of nanocomposite B-MFI-alumina hollow fibre membranes and application to xylene isomer separation, *Micropor. Mesopor. Mater.*, 133, (2010), 18-26.

[17] Gump, CJ; Tuan, VA; Noble, RD; Falconer, JL. Aromatic Permeation through Crystalline Molecular Sieve Membranes, *Ind. Eng. Chem. Res.*, 2001, 40, 565 -577.

[18] Kalipcilar, H; Gade, SK; Noble, RD; Falconer, JL. Synthesis and separation properties of B-ZSM-5 zeolite membranes on monolith supports, *J. Membr. Sci.*, 210, (2002), 113-127.

[19] Bowen, TC; Kalipcilar, H; Falconer, JL; Noble, RD. Separation of C_4 and C_6 isomer mixtures and alcohol-water solutions by monolith supported B-ZSM-5 membranes, *Desalination*, 147, (2002), 331-332.

[20] Ngokoli-Kekele, P; Springuel-Huet, MA; Man, PP; Thoret, J; Fraissard, J; Corbin, DR. ^{128}Xe NMR study of borosilicalites: boron content dependence of xenon–surface interactions, *Micropor. Mesopor. Mater.*, 25, (1998), 35-41.

[21] Axont, SA; Klinowski, J. Solid-State NMR Studies of Zeolite [Si,B]-ZSM-5 Synthesized by the "Fluoride Method", *J. Phys. Chem.*, 98, (1994), 1929-1932.

[22] Miachon, S; Kumakiri, I; Ciavarella, P; van Dyk, L; Fiaty, K; Schuurman, Y; Dalmon, J-A. Nanocomposite MFI-alumina membranes via pore-plugging synthesis: preparation and morphological characterisation, *J. Membr. Sci.*, 281, (2006),228 - 238.

[23] Li, Y; Pera-Titus, M; Xiong, G; Yang, W; Landrivon, E; Miachon, S; Dalmon, J-A. Nanocomposite MFI-alumina membranes via pore-plugging synthesis:Genesis of the zeolite material, *J. Membr. Sci.* 325, (2008), 973-981.

[24] Deng, Z; Nicolas, C-H; Daramola, MO; Sublet, J; Schiestel, Th; Burger, AJ; Guo, Y; Giroir-Fendler, A; Pera-Titus, M. Nanocomposite MFI-alumina hollow fibre membranes prepared via pore-plugging synthesis: Influence of the porous structure of hollow fibres on the gas/vapour separation performance, *J. Membr. Sci.*, 364, (2010), 1-8.

[25] Wegner, K; Dong, J; Lin, YS. Polycrystalline MFI zeolite membranes: xylene pervaporation and its implication on membrane microstructure, *J. Membr. Sci.*, 158, (1999), 17-27.

[26] Chen, HL; Wu, LG; Tan, J; Zhu, CL. PVA membrane filled β-cyclodextrin for separation of isomeric xylenes by pervaporation, *Chem. Eng. J.*, 78, (2000),159-164.

[27] Cunha, VS; Paredes, MLL; Borges, CP; Habert, AC; Nobrega, R. Removal of aromatics from multicomponent organic mixtures by pervaporation using polyurethane membranes: experimental and modeling, *J. Membr. Sci.* 206, (2002), 277-299.

[28] Daramola, MO; Burger, AJ; Pera-Titus, M; Giroir-Fendler, A; Miachon, S; Dalmon, JA; Lorenzen, L. Separation and isomerization of xylenes using zeolite membranes: a short overview, *Asia Pac. J. Chem. Eng.* 5, (2010), 815-837.

[29] O'Brien-Abraham, J; Lin, YS. Effect of Isomorphous Metal Substitution in Zeolite Framework on Pervaporation Xylene-Separation Performance of MFI-Type Zeolite Membranes, *Ind. Eng. Chem. Res.*, 2010, 49, 809-816.

[30] Matsufuji, T; Nishiyama, N; Matsukata, M; Ueyama, K. Separation of butane and xylene isomers with Silicalite-1 zeolitic membrane synthesized by vapour phase transport method, *J. Membr. Sci.*, 178, (2000), 25-34.

[31] Yeong, YF; Abdullah, AZ; Ahmad, AL; Bhatia, S. Process optimization studies of p-Xylene separation from binary xylene mixture over silicalite-1 membrane using response surface methodology, *J. Membr. Sci.*, 341, (2009), 96-108.

[32] Daramola, MO; Burger, AJ; Pera-Titus, M; Giroir-Fendler, A; Miachon, S; Lorenzen, L; Dalmon, JA. Nanocomposite MFI-ceramic hollow fiber membranes via pore-plugging synthesis: Prospects for separation of xylene isomers, *J. Membr. Sci.*, 337, (2009), 106-112.

[33] Daramola, MO; Burger, AJ; Pera-Titus, M; Giroir-Fendler, A; Lorenzen, L; Dalmon, J-A. Xylene vapour mixture separator in nanocomposite MFI-alumina tubular membranes: influence of operating variables, *Sep. Sci. Tech.*, 45, (2010), 21-27.

[34] Xomeritakis, G; Tsapatsis, M. Permeation of Aromatic Isomer Vapors through Oriented MFI-Type Membranes Made by Secondary Growth, *Chem. Mater.*, 11, (1999), 875-878.

[35] Lai, Z; Bonilla, G; Diaz, I; Nery, JG; Sujaoti, K; Amat, MA; Kokkoli, E; Terasaki, O; Thompson, RW; Tsapatsis, M; Vlachos, DG. Microstructural Optimization of a Zeolite Membrane for Organic Vapor Separation, *Science*, 300, (2003), 456-460.

[36] Choi, J; Jeong, HK; Snyder, MA; Stoeger, JA; Masel, RI; Tsapatsis, M. Grain Boundary Defect Elimination in a Zeolite Membrane by Rapid Thermal Processing, *Science*, 325, (2009), 590-593.

[37] Deng, Z; Pera-Titus, M; Guo, Y; Giroir-Fendler, A. Molecular sieving separation of hexane isomers within nanocomposite, (B),-MFI-alumina hollow fibre membranes: a modelling study, *Ind. Eng. Chem. Res.*, 49, (2010), 11697-11707.

[38] Deng, Z; Nicolas, C-H; Guo, Y; Giroir-Fendler, A; Pera-Titus, M. Isomorphously substituted B-MFI hollow fibre membranes for p-xylene separation from C_8 aromatic mixtures, *Sep. Purif. Technol.*, 80, (2011), 323-329.

[39] Bowen, TC; Kalipcilar, H; Falconer, JL; Noble, RD. Pervaporation of organic/water mixtures through B-ZSM-5 zeolite membranes on monolith supports, *J. Membr. Sci.*, 215, (2003), 235-247.

[40] Zhou, R-F; Kong, Y-X; Zhu, M-H; Zhang, F; Zhang, X-L; Chen, X-S; Studies on Synthesis and Pervaporation Performance of Boron-Substituted MFI Zeolite Membranes, Chinese. *J. Inorg. Chem.*, 28, (2012), 942-948.

[41] Nicolas, C-H; Sublet, J; Schuurman, Y; Pera-Titus, M. Role of adsorption and diffusion pathways on the CO_2/N_2 separation performance of nanocomposite, (B),-MFI-alumina membranes, *Chem. Eng. Sci.*, 66, (2011), 6057- 6068.

In: Xylenes
Editor: Michael Olawale Daramola
ISBN: 978-1-62808-342-2

Chapter 8

Thermal Decomposition of Xylene for the Synthesis of Carbon Nanomaterials

D. He, A. Dichiara, H. Li and J. Bai
Laboratoire de Mécanique des Sols, Structures et Matériaux, CNRS UMR 8579, École Centrale Paris, Grande Voie des Vignes, Châtenay-Malabry, France

Abstract

Xylene is an aromatic *hydrocarbon commonly used* in industry as a solvent, but it is also one of the most widely utilized precursors for large-scale production of carbon nanostructures, such as nanotubes, carbon nanofibers or graphene. The carbon feed mechanisms strongly influence the morphology and the quality of these materials. To improve the yield and allow a better control of the synthesis process, a thorough understanding of the chemical kinetics is needed. Hence, the thermal decomposition of xylene was experimentally and numerically studied at temperatures ranging between 500 and 900°C and under atmospheric pressure. The products and kinetics of the xylene pyrolysis were in situ investigated by mass spectrometry and gas phase chromatography and a detailed mechanism was postulated to account for the main products. The evidence of synergistic effects on the xylene decomposition due to the addition of hydrogen and/or acetylene was demonstrated. These results were then applied to the synthesis of various multi-scale hybrid structures for a wide range of applications.

Keywords: Decomposition, carbon, nanomaterials, xylene

Introduction

Xylene is an aromatic *hydrocarbon* commonly produced along with benzene and toluene by extraction and distillation in the petroleum refining process known as catalytic reforming. It can be also created from coal carbonisation in the manufacture of coke fuel and more recently by *new synthesis routes involving alternative and renewable feedstocks [1,2]. Xylene* encompasses three isomers of dimethylbenzene, distinguished by the designations ortho-

(o-), meta- (m-), and para- (p-), specifying to which carbon atoms (of the benzene ring) the two methyl groups are attached. Some chemical and physical properties differ from isomer to isomer, for instance, the melting point −47.87°C for m-xylene) while 13.26°C for p-xylene. The boiling point for each isomer is about 140°C. The density of each is around 0.87 g ml^{-1}, less dense than water.

Xylene has a wide range of applications in various industrial fields. Hence, p-xylene is the principal precursor for terephthalic acid and dimethyl terephthalate, both monomers used in the production of polyethylene terephthalate plastic bottles and polyester clothing. Xylene is a widely used solvent, where the mixture of isomers is often referred to as xylenes or xylol. Like the individual isomers, the mixture is colorless, sweet-smelling, and highly flammable. Areas of application include printing, rubber, and leather industries. In addition, it is a cleaning agent, e.g., for steel, silicon wafers, and integrated circuits. In dentistry, xylene can be used to dissolve gutta percha, a material used for endodontics (root canal treatments). In the petroleum industry, xylene is also a frequent component of paraffin solvents, used when the tubing becomes clogged with paraffin wax.

Xylene as a Precursor for the Synthesis of Nanostructures

While xylene is widely used as solvent in the adhesive, pesticide, pharmaceutical, printing and rubber industries, it is also utilized as a raw material and chemical intermediate for various other applications. An important *one* is to serve as carbon precursor for the production of carbon nanomaterials for example, carbon nanotubes (CNTs), carbon nanofibers and graphene. The CNTs have attracted worldwide attention in the last decades. They may have many potential applications in a wide range of fields such as aerospace, energy storage, biotechnology and polymer composites [3,4]. This is mainly due to their light weight, specific structures and extraordinary mechanical [5-7], thermal [8,9] and electrical [10,11] properties. However, their properties are strongly structure-dependent. A CNT could be seen as a curved graphene cylinder rolled-up along a certain axis. According to the number of curved graphene sheets, CNT is commonly classified into single-walled CNT (SWCNT), double-walled CNT (DWCNT) or multi-walled CNT (MWCNT). At the same time, a SWCNT could also be classified into armchair, zigzag or chiral type according to their rolling way (i.e. their chirality). Consequently, their electrical property varies from semi-conducting to metallic [12]. Meanwhile, different graphitization degrees of CNTs could considerably influence their mechanical properties, electrical and thermal conductivities, as well as their chemical stability.

During the last two decades, a large variety of techniques have been explored to produce CNTs, including electric arc discharge [13-17], laser ablation [18], chemical vapor deposition (CVD) [19-22], ball milling [23], diffusion flame [24-26], electrolysis [27-30], use of solar energy [31], heat treatment of a polymer [32], and low-temperature solid pyrolysis [33]. Among them, CVD is the most widely used one, due to its outstanding characteristics like easily scaled up and relatively low operating temperatures. More particularly, a promising way for the CNT synthesis consists in growing the nanostructures onto the desired substrates by simultaneously decomposing a mixture of organometallic and hydrocarbon precursors [34-38]. This method, called floating catalyst CVD, allows low cost and continuous productions [39]. High yield mass production can be achieved by replacing the typical flat substrates by others with larger surface area, like micro-beads.[35,37,40] In a general floating CVD process, three components are usually required for CNT synthesis: carbon source, heat source and catalyst precursor. Xylene is one of the most widely used carbon sources, especially for MWCNT synthesis. Its decomposition kinetics has significant influence on the CNT nucleation, structure and growth rate. The large-scale production of CNTs with homogeneous diameter and structure is particularly desired for their property research and industrial applications.

In this chapter, the thermal decomposition of xylene was experimentally and numerically studied at temperatures ranging between 500 and 900°C under atmospheric pressure and its influence on the CNT growth was examined. First, the influence of the xylene decomposition on the CNT morphology and organization was investigated. The products and chemical kinetics of the xylene pyrolysis were then *in situ* analysed by mass spectrometry (MS) and gas phase chromatography (GPC). The reduced and detailed mechanisms were postulated to account for the main products. The evidence of synergistic effects on the xylene decomposition due to the addition of hydrogen and/or acetylene was established. These results were applied to the CVD synthesis of CNTs on ceramic micro-spherical particles to form hybrid structures that can be easily produced in large scales for advanced multifunctional composite applications.

Experimental Details

A horizontal CVD reactor was used to synthesize CNTs by a floating CVD method as reported in reference [35,37,41,42]. The temperature profiles along the inner wall of the quartz tubes were measured at given temperatures under argon flow (1 l min^{-1}). Briefly, a 1.2 m long quartz tube with an inner diameter of 45 mm was heated by a horizontal tube furnace (Carbolite®). The alumina microparticles (μAl_2O_3, size range 1–5 μm, with 99.8% purity including 800 ppm SiO_2, 600 ppm Na_2O purchased from Performance Ceramics Company) were homogeneously dispersed on the surface of a quartz plate (0.5 m in length) which was then put in the center of the heated region (0.6 m in length). In each experiment, the relative position between the quartz tube and the electrical resistance furnace was fixed. The spray nozzle consisted of two stainless steel tubes. One tube was for the carrier gas ($Ar:H_2:C_2H_2$), while the other one was used to inject the ferrocene/xylene solution (0.05 g ml^{-1}) at 0.2 ml min^{-1}. The nebulization of the solution was formed at the point where it encountered the gas. The droplets were carried forward, and quickly vaporized before arriving at the entrance of heated zone. We considered approximately that all the droplets were transferred into gas at the entrance of the furnace. The injection of both the carbon and catalyst precursors lasted 15 min. After the reactor was cool down to room temperature under argon and hydrogen atmosphere, the powder was separately collected into several small bottles according to their locations in the CVD reactor, and characterized using various techniques such as SEM, TEM and Raman spectroscopy.

The CVD gas phase was analysed by in situ MS (Pfeiffer Vacuum GSD 301 Thermostar) and *in situ* GPC composed of a thermal conductivity detector and a packed column using helium (He) as carrier (Alpha MOS PR2100). Both MS and GPC were equipped with a 90 cm long capillary which was drilled through a rubber plug set, for gas sampling. The gas sampling position was set up few centimeters above the substrate location at the center of the reactor to analyse the gas in the CVD, avoiding the cooling down process. Each MS and GPC result corresponds to an average of 3 measurements. The data were normalized as a function of the Ar intensity.

RESULTS AND DISCUSSIONS

Variation of CNT Structures with Xylene Decomposition Temperature[35,43]

Figure 1 shows the different CNT organizations on μAl_2O_3 and the corresponding consecutive transformations with the increase of CNT diameter. These hybrid structures were prepared using xylene as carbon source at temperatures ranging from 550 to 900°C. The total gas flow rate was fixed to 0.8 l min^{-1}, in which hydrogen and argon flow rates were set up at 0.08 l min^{-1} and 0.72 l min^{-1}, respectively. Vertically aligned highly dense MWCNTs grown along six different directions on μAl_2O_3 at 550°C form the so-called "six-branch" hybrid structures (Figure 1a). The CNT diameters and lengths are ~ 10 nm and ~1 μm, respectively. When CNT diameters are around 10 nm (600°C) or 15 nm (650°C), they are always configured in "six-branch" (Figure 1b-c). However, the "six-branch" structures become less evident when the CNT average diameters increase to ~20 nm at 700°C (Figure 1d). Beyond this temperature, the chemical decomposition of xylene is intensified. Finally, both the average diameter and length increase continuously with the temperature, while the area number density decreases. As a result, the "urchin-like" hybrid architecture appears at 750 and 800°C (Figure 1e-f). The CNT diameters have a considerable augmentation and vary in a large region at these temperatures. Indeed, SEM and TEM observations confirm a decrease of CNT area number density on the surface of μAl_2O_3. Both diameter and length of CNTs increase continuously with the temperature, and even drastically after 700°C. The CNT aspect ratio fluctuates in a narrow range from 100 to 300.

A further increase of the temperature results in much longer and larger CNTs, ~150 μm in length and ~50-100 nm in diameter at 850°C (Figure 1g). Whereas, by comparing with the hybrid structures obtained at lower temperatures, the number density of CNTs on μAl_2O_3 is greatly decreased. Moreover, CNT arrangement on the substrate surface also changes from a vertically aligned state to a random one. Often, it is found that several small μAl_2O_3 are attached to a larger one. However, the hybrid structures are completely changed when the temperature is increased to 900°C. Only few CNTs decorated with iron particles are formed on μAl_2O_3 (Figure 1h). Instead, the higher magnification SEM (Figure 1i) of one selected region on the surface of μAl_2O_3 and its corresponding secondary electron image (Figure 1j) show that μAl_2O_3 are densely covered by amorphous carbon balls. Each ball encapsulates a small iron particle sized in a range from 10 to 20 nm. It is worth mentioning that the CNT diameter is similar to the iron particle size, and no CNT with large diameter is found. This result is coherent with that reported by Eres et al. [44], that is, the CNT growth with ferrocene losses vertical alignment at 900°C, and the growth rate exhibits a dramatic steep drop. This is mainly due to the enhanced gas-phase thermal decomposition of $Fe(C_5H_5)_2$ above 850°C, which immediately produces iron particles in gas phase before ferrocene molecule arrival on the substrate surface. The enhanced thermal decomposition of carbon sources at high temperatures could also accelerate the carbon deposition on the forming iron particles. One direct consequence is the presence of a large amount of carbon encapsulated iron particles on the outside walls of CNTs and on the μAl_2O_3 surface (Figure 1h-j).

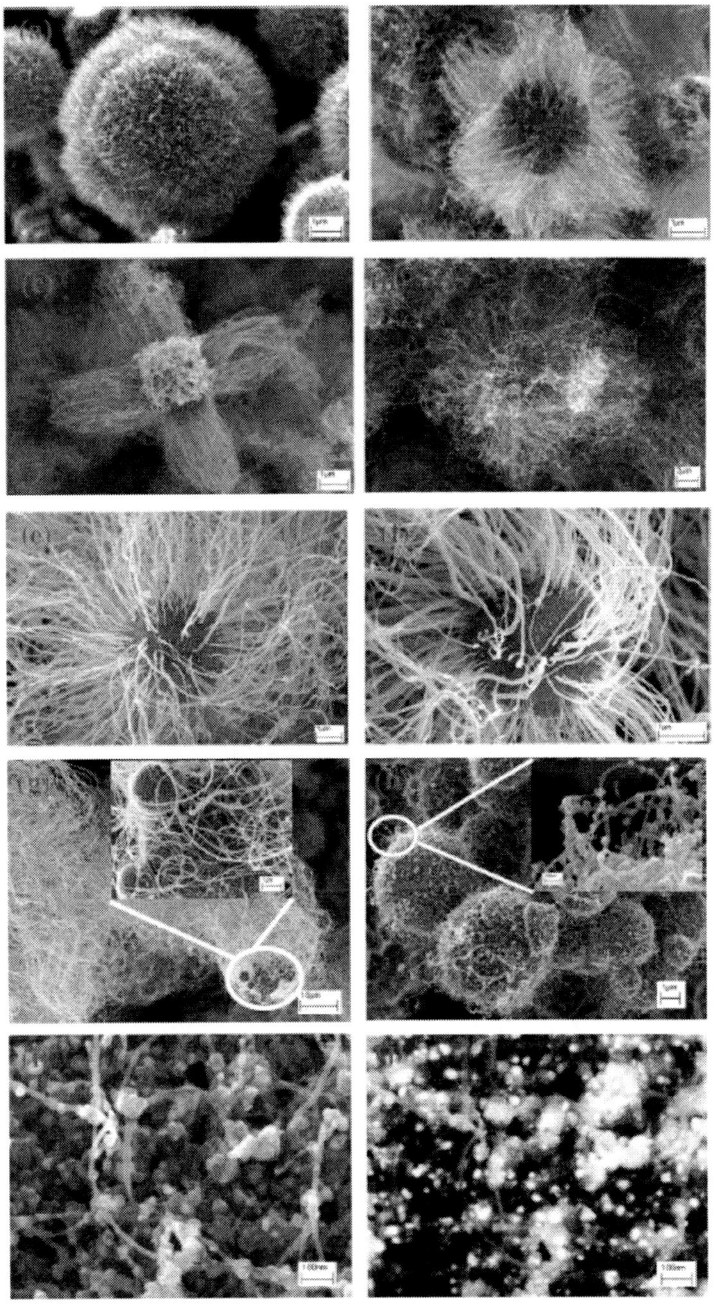

Figure 1. SEM images of the self-organized CNT patterns on μAl_2O_3 obtained at different temperatures: (a) 550°C, (b) 600°C, (c) 650°C, (d) 700°C, (e) 750°C, (f) 800°C, (g) 850°C, and (h-j) 900°C. (i-j) High magnification images of the selected red zone in image (h) [35].

Figure 2a shows the main MS intensity peaks which have evident variations when the temperature increases from 400 to 1000°C. With the temperature increase, the decomposition process of the ferrocene-xylene solution undergoes three different periods, marked by two red dotted lines. The slight augmentation of the curves from 400 to 600°C might be due to the incomplete evaporation of the solution spray droplets at low temperatures. In fact, only a small proportion of xylene decomposes into toluene at this temperature range. This could be seen from the variation of the intensity ratios of I_{106}/I_{91} (Figure 2b) and I_{92}/I_{106} (Figure 2c). The MS of pure o-xylene has the two strongest ion intensity peaks at m/z 91 and 106, which are attributed to the species $C_7H_8^+$ and $C_8H_{10}^+$, respectively. In the MS of toluene, there is also a characteristic peak at m/z 91 corresponding to toluene molecule. In Figure 2b, a slight decrease of the intensity ratio I_{106}/I_{91} indicates a weak decomposition (~3 % at 600°C) of xylene when the temperature is between 400 and 600°C. On the other hand, a slight near linear increase (~2.5 % at 600°C) of I_{92}/I_{106} is found at this temperature range (Figure 2c). Furthermore, no variation of I_{78}/I_{92} could be seen with the increase of the temperature from 400 to 600°C (Figure 2d). This indicates that no benzene is produced at these temperatures. Therefore, the decomposed xylene is almost completely transferred into toluene when the temperature is lower than 600°C.

In the second stage, the decomposition of the xylene solution increases when the temperature varies from 600ºC to 900°C. Figure 3 shows the quasi-linear decreases of the ion intensities at m/z 106 ($C_8H_{10}^+$) and 91 ($C_7H_7^+$), whereas the ones at m/z 78 ($C_6H_6^+$) and 15 (CH_3^+) increase with the temperature. However, the I_{92} ($C_7H_8^+$) first shows an increase from 600 to 800°C, and then a decrease from 800 to 900°C. The evident increase of I_{78} ($C_6H_6^+$) could only be seen after 750°C. Moreover, the intensity ratio I_{106}/I_{91} decreases in a nearly linear way from 0.38 to 0.01 with the increase of the temperature from 600°C to 900°C (Figure 2b). The decreasing I_{106}/I_{91} indicates that there are certainly the generation of C_7H_8 species in gas phase due to the increased decomposition of C_8H_{10}. At the same time, the ratio I_{92}/I_{106} undergoes a firstly slow linear increase from 0.2 at 600°C to 1.85 at 750°C, and then a dramatic one above 750°C (Figure 2c). This means that there remain a few amounts of xylene in the gas phase above 750°C. The further observation (Figure 2d) of the variation of the ion intensities at m/z 106, 92 and 78 shows that most xylene is transferred into toluene from 600 to 750°C. At this temperature region, there is a very weak augmentation of benzene concentration. Whereas, the rapid augmentation of I_{78} indicates the increased production of benzene in a linear way, with the increase of temperature from 750 to 900°C. Benzene could be produced from either the residual xylene, or the generated toluene. As shown in Figure 2d, both the I_{78}/I_{40} and I_{106}/I_{40} decrease with the increase of the temperature. However, the former exhibits a slower decrease rate than the latter. The similar phenomena are also seen from the variations of the relative intensities of I_{106}/I_{78} and I_{92}/I_{78} (Figure 2e). Therefore, it can be deduced that benzene is mostly produced from the decomposition of toluene. At 900°C, the concentration of benzene is significantly higher than this of toluene, which is further greatly higher than that that of xylene (Figure 2c and d). In addition, the ion intensities at m/z=12 (C^+) and 26 ($C_2H_2^+$) have a weak increase with the temperature. Whereas, the intensity at m/z=51 first decreases in the temperature range from 600 to 750°C, and then increases. This is mainly due to the variation of the toluene and xylene concentrations in the gas phase.

In the third temperature region ($T \geq 900$°C), the benzene concentration starts decreasing, and the residual toluene is continuously decomposed. When the temperature is equal to 1000°C, the levels of benzene and toluene in the residual gas are very weak, almost zero.

However, there remains a little of xylene in the reactor atmosphere. This results in an increase of the ratio I_{106}/I_{91} (Figure 2b) and a decrease I_{92}/I_{106} (Figure 2c). Moreover, the species CH_3 and CH_4 formed from the decomposition of xylene and toluene are largely consumed at 1000°C.

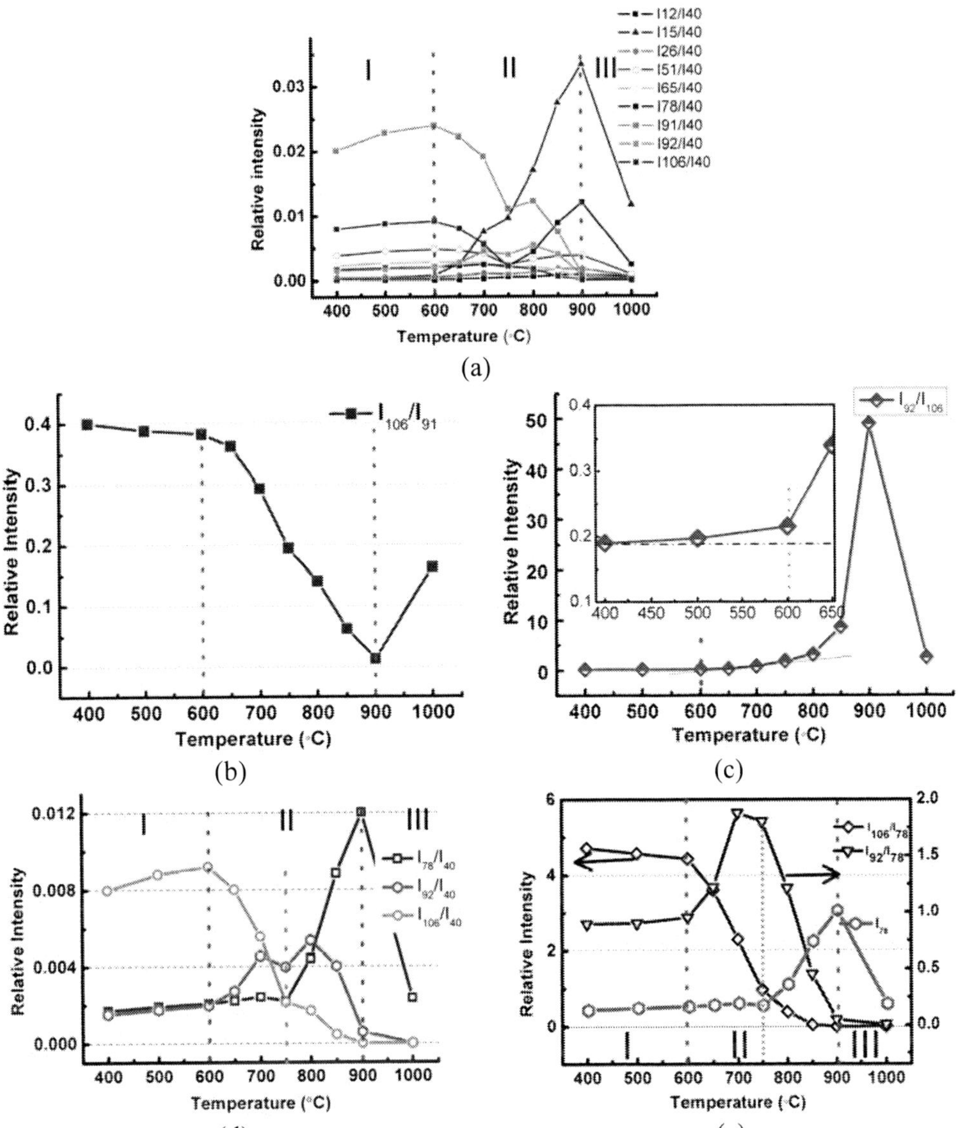

Figure 2. (a) MS analysis of the decomposition of xylene solution containing 0.05 g ml^{-1} ferrocene at the temperatures ranging from 400 to 1000°C. Two red dot lines indicate two critical temperatures of the decomposition of the solution. (b) The ratios between the ion intensities at m/z=106 ($C_8H_{10}^+$) and m/z=91($C_7H_7^+$). (c) The ratios between the ion intensities at m/z=92 ($C_7H_8^+$) and m/z=106 ($C_8H_{10}^+$). The inserted figure is the amplification of the curve between 400 and 650°C. (d) The relative ratios between the ion intensities at m/z=106 ($C_8H_{10}^+$), m/z=92 ($C_7H_8^+$), m/z=78 ($C_6H_6^+$) and m/z=40 (Ar^+). (e) The relative intensity ratios of I_{106}/I_{78} and I_{92}/I_{78}. For comparison, the ion intensity curve of I_{78}/I_{40} is plotted at the bottom of figure.[43]

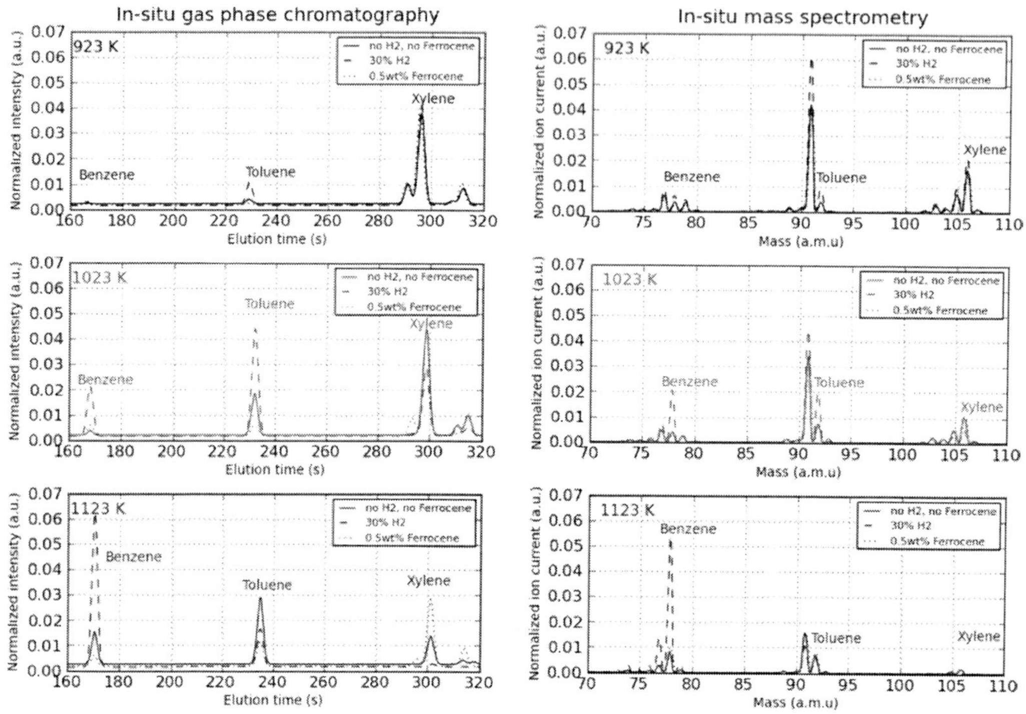

Figure 3. In situ GPC (left column) and MS (right column) analyses during the pyrolysis of C_8H_{10} at different temperatures without H_2 addition (straight line), with 30 vol.% H_2 (dashed line) or with 0.5 wt.% ferrocene (dotted line) [37].

Therefore, it can be seen that the xylene solution undergoes a complex decomposition process when the temperature varies in the range from 400 to 1000°C. The whole temperature region could be divided into 3 major zones. In the first region ($T \leq 600°C$), there is only a little of xylene decomposed under the used conditions. As a result, short CNTs with small diameter are grown on mAl_2O_3 at a slow rate. In the second region ($600°C \leq T \leq 900°C$), the increased pyrolysis rates of xylene and its subsequent products such as toluene promote the growth of CNTs with large diameters and lengths. However, the high decomposition rates at 900°C also promote the direct formation of amorphous carbon in gas phase. Beyond this temperature, no CNTs are grown on mAl_2O_3 under the investigated conditions.

Influence of Hydrogen on Xylene Decomposition

Certainly high synthesis temperatures are required for xylene pyrolysis due to the strong C-C bond strength in aromatic rings. However, high reaction temperatures were found to result in low CNT production rates and high concentration of amorphous carbon residue in the reactor, as reported above. Therefore, decreasing the temperature of xylene pyrolysis becomes of paramount importance for the mass production of CNTs. To this aim, certain amounts of hydrogen (~ 10%) can be added into the carrier gas during the catalytic synthesis of CNTs [37]. The addition of hydrogen can also prevent the oxidation of catalyst particles as well as the formation of carbon impurities during CNT growth. Hence, at a given temperature, the *in situ* gas phase analysis showed that when H_2 is added in the reacting atmosphere, the level of C_8H_{10} drops off while these of its byproducts increase. This indicates that the pyrolytic dissociation of C_8H_{10} is significantly accelerated in the presence of H_2, as depicted in Figure 3. H_2 addition-exchange reactions being much easier than the direct bond cleavage reactions, C_8H_{10} and further C_7H_8 can decompose into C_6H_6 by hydrogenolysis as suggested by Benson et al. [45]. However, when the H_2 flow rate becomes too high, it may induce suppressing effects on the carbon structure synthesis. For instance, H_2 rates higher than 30 vol.% were found to impede the CNT growth by driving the reverse reaction for decomposition due to the gasification of the deposited carbon [46], etching sp2 bonds [47], inducing catalyst deactivation by Otswald ripening [48,49] and saturating dangling carbon bonds [50]. Therefore, optimal amount of H_2 catalyses the decomposition of xylene and allows to lower the reaction temperature. Nevertheless, the xylene decomposition remains very weak below 600°C even in the presence of H_2. Therefore, the use of an additional carbon source is needed to allow the growth of CNTs at lower temperatures.

Figure 4. SEM images of CNTs-Al_2O_3 hybrid structures obtained using 20 ml min^{-1} C_2H_2 and 0.2 ml min^{-1} ferrocene-xylene solution at the following temperatures: (a) 475°C, (b) 500°C, (c) 550°C, (d-e) 600°C, (f-g) 650°C and (h-i) 700°C. Images (e), (g) and (i) are SEM images of the hybrid structures collected in the latter regions of the quartz plate at the temperatures given. The CNT bundle length at each temperature is indicated in the corresponding figures. The inserted histograph in (i) shows the diameter distribution of CNTs in the hybrid structures [51].

Use of Acetylene as an Additional Carbon Source and Its Interaction with Xylene [51]

As previously detailed, additional carbon feedstock(s) can be used to lower the CNT synthesis temperature. Hence, the effect of temperature is studied in the range from 450 to 700°C when the xylene-acetylene mixture is used as carbon source. Ferrocene, concentrated at 0.05g ml^{-1}, was dissolved in xylene, which was fed into the furnace at a rate of ~ 0.2 ml min^{-1}. In the meantime, C_2H_2 was fed at a flow rate of 20 ml min^{-1}. The feeding of the ferrocene-xylene solution and acetylene lasted for 15 min. Argon and hydrogen served as carrier gases at flow rates of 0.88 and 0.1 l min^{-1}, respectively.

The influence of acetylene on the CNT growth on μAl_2O_3 is greatly different at low and high temperature ranges. This is mainly due to varied thermal decomposition mechanism of acetylene and xylene with the reaction temperature. Acetylene plays different roles in the CNT growth at different temperatures. In the following part, we discuss the decomposition of xylene/acetylene mixture in the temperature range from 450 to 700°C when 2% acetylene is added into the reactor.

Figure 4 displays SEM images of the CNTs-μAl_2O_3 hybrid structures obtained at the temperature ranging from 475 to 700°C. The hybrid structures consisting of six CNT bundles are abundantly observed in the final products when the temperature is beyond 450°C (Figure 4a-h). At low temperature from 475 to 550°C, the CNT bundles are regularly distributed in the six orthogonal directions. The CNT length consecutively increases from ~1 μm at 475°C to ~9 μm at 550°C, but CNT diameter remains in the range of 5-10 nm. The CNT bundles obtained at 550°C (Figure 4c) are also quite different from these at lower temperatures. The CNTs in the bundles obtained at 550°C are nearly parallel to each other. Whereas the CNT bundles obtained at the lower temperatures are contracted at each end. High density CNTs are observed on the alumina particles placed in the quartz plate between 5 and 20 cm at 600°C (Figure 4d). The CNT length is ~20 μm and its diameter remains ~10 nm. However, CNTs grown on μAl_2O_3 beyond 30 cm is much shorter in length (~8 μm) but larger in diameter ranging between 10 -15 nm (Figure 4e). Similarly, the long CNT bundles around 45 μm are found on μAl_2O_3 located between 5 and 20 cm at 650°C (Figure 4f), but shorter bundles on the particles located in the range of 20-30 cm (Figure 4g). However, the diameter of CNTs in the different regions has little variation, ranging between 10 -15 nm. Beyond this temperature, the CNT density evidently decreases on the alumina microparticles which are located at the first 10 cm of the quartz plate (Figure 4h). The bundles present a reduced length, ~25 μm. Moreover, a variation of CNT diameter is also observed from 5 to 15 nm. Most of CNTs have diameters ranging from 10 to 15 nm. No CNT bundle could be observed on the microparticles located in the latter region of the reactor (Figure 4i). The CNTs of 10 μm in length arrange in all directions on the alumina particles, forming the "urchin-like" hybrids. The diameter distribution bar diagram (inserted in Figure 4i) shows that most CNTs have diameters ranging from 15 to 25 nm, and a few CNTs having a big diameter of 40 nm, or even 60 nm.

Figure 5 shows the distribution of CNTs obtained at the temperatures ranging from 500 to 700°C on Al_2O_3 which are homogeneously dispersed on the surface of a 50 cm long quartz plate. It could be seen that the starting deposition position of CNTs moves toward the front of quartz plate with the increase of temperature. The decomposition of xylene becomes intense after 600°C. This could explain why the CNT diameter largely increases after this temperature, especially at 700°C. At the same time, the highly dense CNT deposition region narrows with the increase of temperature above 550 °C. This is mainly due to the increased consumption rates of ferrocene at higher temperatures. The following MS measurements give more explanations.

Figure 6a shows the mass spectra of the exhaust gas produced by the decomposition of the acetylene and xylene mixture at 550°C. The ion current peaks at m/z 40(20) and 2 are attributed to the carrier gases Ar and H_2, respectively. The ones at m/z 44, 32, 28 and 14 correspond to the components of residual air such as CO_2, O_2 and N_2, respectively. Acetylene and xylene are identified by their specific molecular peaks at m/z 26 and 106, respectively. The fragments C_4H_x, C_5H_x, C_6H_x, C_7H_x (subscript x indicates integral numbers) resulting from the pyrolysis in furnace or the ionizations in mass spectrometer are displayed in the spectra.

Figure 6b shows the evolution of the ion currents at the masses corresponding to $C_8H_{10}^+$ (m/z 106), $C_7H_8^+$ (m/z 92), $C_6H_6^+$ (m/z 78), $C_5H_5^+$ (m/z 65), $C_4H_{10}^+$ (m/z 58), $C_4H_4^+$ (m/z 52), $C_2H_2^+$ (m/z 26) and CH_4^+ (m/z 16) with the increase of temperature from 450 to 900°C. Some noticeable intensity changes are observed at m/z 16, 26, 78, 92 and 106, which correspond to the fragments of CH_4^+, $C_2H_2^+$, $C_6H_6^+$, $C_7H_8^+$ and $C_8H_{10}^+$, respectively. Obviously, $C_2H_2^+$ undergoes a significant change: it first decreases very slowly from 450 to 550°C, and then abruptly to a constant at 700°C. The ion current at m/z 106 corresponding to xylene exhibits first a slight increase until 550°C due to the increasing evaporation of the xylene solution. Then, the current starts decreasing because of the beginning of the decomposition. The intensity at m/z 92 undergoes an important increase from 550 to 800°C, and then an evident decrease after 800°C. This indicates that toluene begins to largely decompose after this temperature. However, the monotonous increase of the relative intensity at m/z 78 indicates the continuous generation of benzene in the reactor, especially after 700°C. The intensity at m/z 66 corresponding to the fragment of $C_5H_6^+$ starts to decrease significantly only when the temperature becomes higher than 800°C.

Furthermore, the decomposition of xylene and the generation of toluene and benzene can be more precisely evaluated from the evolution of the relative ratios between the intensities at m/z 106, 92, 91, 78 and 77. According to the standard spectra library provided by Pfeiffer Vacuum, the relative intensity ratios are different in the spectra of xylene, toluene and benzene. They are as follows: I_{106} / I_{91}: ~0.577 (xylene), I_{92} / I_{91}: ~0.075 (xylene) and ~0.784 (toluene), and I_{78} / I_{77}: ~0.661 (xylene) and ~6.938 (benzene). The evolutions of the I_{106} / I_{91}, I_{92} / I_{91} and I_{78} / I_{77} (Figure 6c) show that xylene does not evidently decompose under these experimental conditions until 550°C. The decomposition rate of xylene increases quickly after 600°C. Similarly, it is found that toluene and benzene appear in exhaust gas obtained at temperatures higher than 550°C. Their concentrations quickly increase, meaning that a large part of xylene decomposes into toluene and benzene.

The impact of acetylene on the decomposition of xylene is further demonstrated by comparing the relative intensities at m/z 78 ($C_6H_6^+$), 92 ($C_7H_8^+$) and 106 ($C_8H_{10}^+$) between the two cases with or without acetylene. MS measurements are conducted on the following two kinds of exhaust gases at the temperatures ranging from 450 to 900°C. In the first case, xylene is used as the only carbon source. In the second case, acetylene is added into the reactor at a flow rate of 20 ml min^{-1}. The concentration of ferrocene in xylene solution is 0.05 g ml^{-1}. As shown in Figure 7a, the relative intensity of xylene with C_2H_2 is lower than that without C_2H_2 when the temperature is lower than 700°C. However, no evident difference has been observed in the relative intensities at m/z 78 and 92 between the cases with or without addition of acetylene. Beyond 700°C, the addition of acetylene produces little variation of the intensities at m/z 92 and 106, but an increasing augmentation of the one at m/z 78. This means that more acetylene is transformed into benzene at high temperatures than at low ones. As shown in Figure 7b, the intensity ratios I_{106}/I_{78} and I_{92}/I_{78} of two cases (with and ~~no~~ without C_2H_2) also indicate that the addition of acetylene generates a slight decrease of the relative abundances of xylene and toluene in the system, compared with that of benzene.

Based on the MS analyses and SEM characterizations, it could be seen that the evolution of the hybrid structures is mainly related to the pyrolysis of C_2H_2 at the temperatures ranging from 450 to 600°C. The synthesized hybrid structures consist of CNTs in "the six-branch" organization, and the CNTs have a homogeneous diameter, ~5-10 nm. A further increase of the temperature yields the increased decomposition of xylene, which contributes more carbon containing fragments to CNT growth. Due to the intensified decomposition of ferrocene when the temperature is higher than 600°C, CNT diameter and length increase with the temperature. The "urchin-like" hybrid structures which consist of CNTs with large diameters are synthesized at temperatures higher than 700°C.

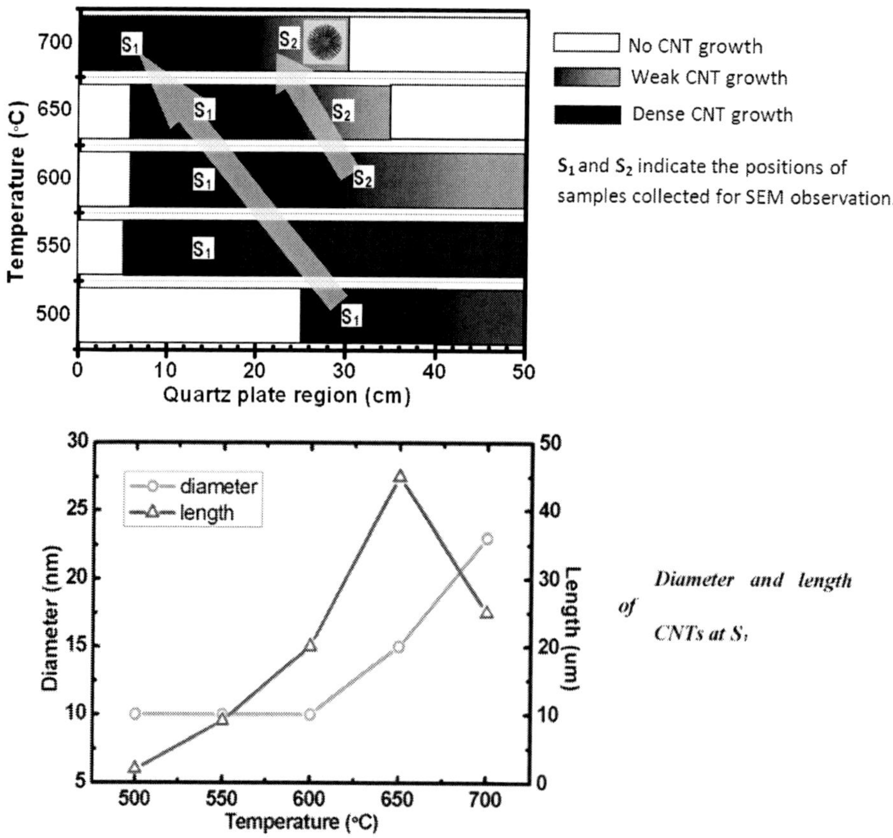

Figure 5. Schematic of CNT deposition on alumina microparticles dispersed on quartz plate at different growth temperatures. The color change represents the variation of CNT deposition quantity. Two yellow arrows show the increasing tendency of CNT diameter. The figure at right bottom shows the variations of CNT diameter and length with the temperature [51].

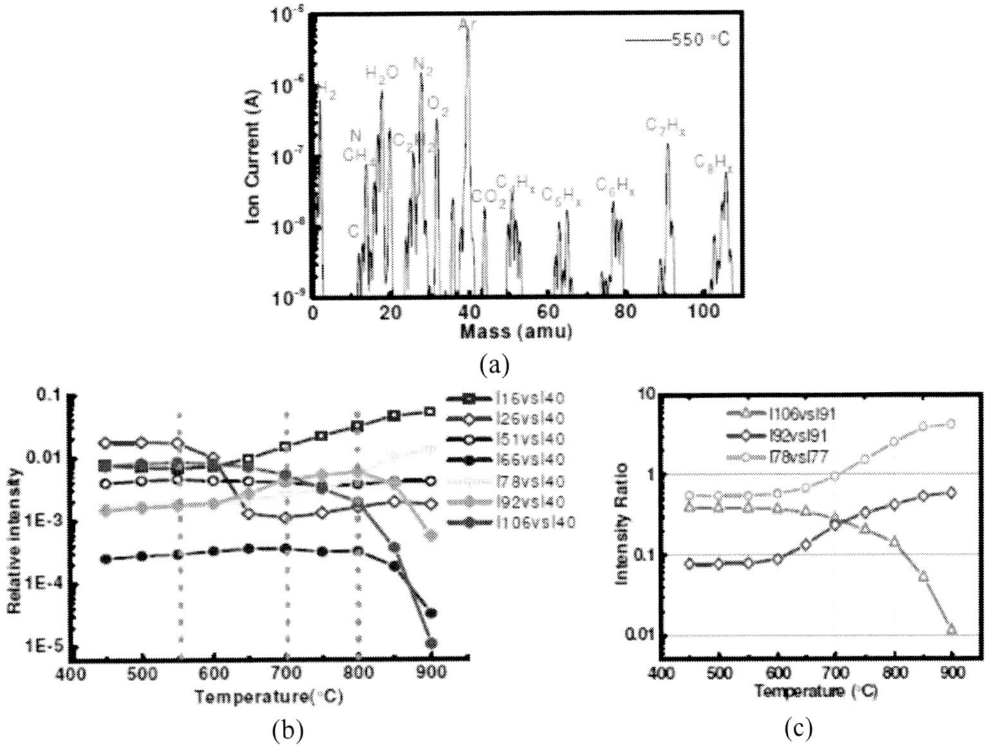

Figure 6. (a) Mass spectra of the exhaust gas at 550°C. (b) Evolution of the ion currents at masses (m/z) corresponding to the predominant fragments formed during the reactions. (c) Relative intensity ratios of the fragments of C_6H_5, C_6H_6, C_7H_7, C_7H_8, C_8H_{10}, showing the decomposition of xylene and the production of benzene.[51]

Interaction Effect between Acetylene and Xylene at Higher Temperature

The previous discussions of CNTs grown from different carbon sources have shown notable differences in CNT diameter, length and morphology, due to the variations of the chemical reaction process. From the CNT amount increasing extent, we notice the increasing CNT growth efficiency when acetylene is used. It is therefore believed that a strong interaction exists between the decomposition of xylene and acetylene. The decomposition rate of xylene, toluene and benzene at different acetylene flow rates was determined at 800°C, as shown in Figure 8 [42]. Obviously, the decomposition of xylene is intensively promoted by the addition of acetylene. For the reaction process from xylene to toluene, the decomposition rate (r_1) increases from 83.3 % to 97.2 %, indicating at the experimental temperature, xylene is sufficiently decomposed. However, when only xylene is used as carbon source, benzene concentration in the effluent gas is 1.67 %. When 50 ml min^{-1} of acetylene is introduced, this value decreases to 0.57 %. This phenomenon is a competitive result of the interaction between various factors. First, acetylene can be polymerized to benzene [52, 53] ~~in~~ under the applied experimental condition. T.C. Schmitt [54] proposed a series of hydrocarbon products generated from acetylene in the catalytic CVD process of CNTs, including benzene and cycloocta-1,3,5,7-tetraene etc. During our measurement, we can also clearly identify the peaks for benzene in the mass spectrum when acetylene is the only carbon source. This reaction will increase the benzene concentration in the gas phase, and restrain the generation of benzene from xylene decomposition process by increasing the product concentration in the chemical equilibrium. On the other hand, acetylene will react with toluene and benzene to form a series of byproducts as shown both in the detailed mechanism model and in the MS measurement. This will consume benzene and toluene in the gas phase, hence decreases the benzene concentration and promotes the decomposition of xylene. When acetylene is introduced, the decomposition rate of toluene and benzene both increase intensively from 41.6 % to 83.8 % (r_2) and from 7.6 % to 80.9 % (r_3), which indicates the reaction consuming toluene and benzene contribute more than the reaction producing benzene.

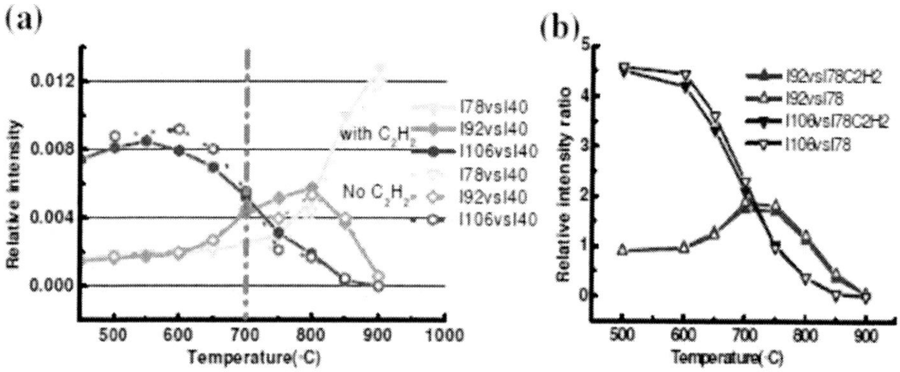

Figure 7. MS analyses of the impact of acetylene on the decomposition of xylene into toluene and benzene. (a) Comparison of relative intensities at m/z 78 ($C_6H_6^+$), 92 ($C_7H_8^+$) and 106 ($C_8H_{10}^+$) obtained by measuring of the exhaust gases from the decomposition of carbon source with (solid symbols) or ~~no~~ without (hollow symbols) acetylene. (b) Relative intensity ratios I_{106}/I_{78} and I_{92}/I_{78} in the cases with (solid symbols) or without acetylene (hollow symbols) in the system [51].

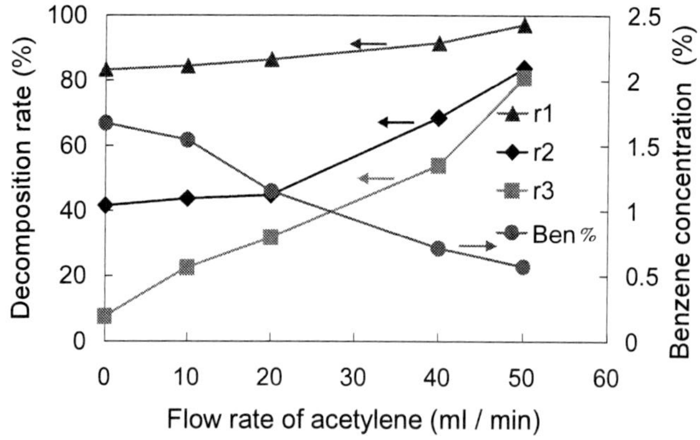

Figure 8. Variation of the decomposition rates and benzene concentration with increasing acetylene flow rate. r_1, r_2 and r_3 denote the decomposition rate of xylene, toluene and benzene, respectively. Ben % denotes benzene concentration in the effluent gas [42].

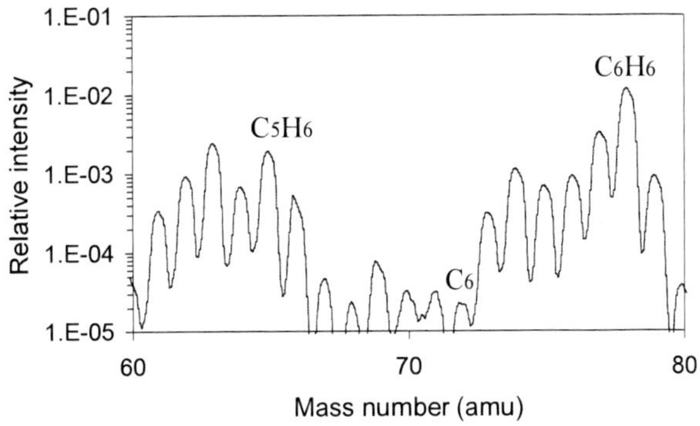

Figure 9. Mass spectrum of gaseous components with logarithm coordinates. C_6: Peak for the hexagonal carbon rings on mass number 72 amu [42].

From the mass spectrum, we notice that the proportion of naphthalene is higher than other hydrocarbon byproducts, meaning that the amount of benzene consumed is more than that of toluene consumed. So, the promotion effect of acetylene to toluene is higher than that to xylene. When acetylene flow rate increases from 0 to 50 ml min^{-1}, the decomposition rate of toluene (r_2) increases by 101.4 %, and the decomposition rate of xylene (r_3) increases by 16.1 %. This effect can provide more effective carbon feedstocks for the CNT growth, which intensively increases the CNT growth rate. However, the amount of benzene consumed by acetylene is much lower than the amount of benzene decomposed. Interestingly, no large amount of other hydrocarbon products is detected by MS. However, the trace peak on mass number 72 amu in the mass spectrum (Figure 9) with logarithm coordinates provides valuable information. By comparing with the standard mass spectrum, the existence of other species, like pentane, neopentane, 2-methylbutane, that may have peaks on mass number 72 amu are rejected. On the contrary, H_2 generated from this process is released to the gaseous atmosphere. It can be easily detected by MS on mass number 2 amu.

Based on the previous results and discussions, two different cases can be identified. First, the CVD synthesis at relatively low temperatures where either hydrogen, or acetylene, or both are needed to grow CNTs. In this case, while xylene has been found to be able to decompose into toluene and benzene by hydrogen addition-exchange reactions, the evolution of the hybrid structures is mainly related to the pyrolysis of C_2H_2 when it is added into the reactor. The second case consists of the high speed CNT growth due to the intensified xylene decomposition, and their interactions with acetylene at high temperatures. In the next part of this chapter, the chemical gas phase reactions will be studied numerically.

Gas Phase Chemical Reaction Analyses: Experimental and Numerical Approaches

Chemical Reaction Kinetics at Relatively Low Temperatures

Here, we will focus on the main gas-phase chemical reactions occurring in the reactor at temperature ranging between 550 and 780°C. At such temperatures, xylene decomposes principally into toluene and benzene by hydrogen addition-exchange reactions.

Decomposition of Xylene

Due to the strong R-CH_3 (R represents aromatic ring) bonds in toluene and xylene (434.7 kJ mol^{-1}) [55], the direct transformations of xylene and toluene to toluene and benzene by R-CH_3 cleavage are negligible between 450 and 780°C, respectively. Instead, the hydrogen (H) addition-exchange reactions are much easier. Benson and Shaw [45] set up the following two global reactions by combining the H substitution reactions with an indirectly set-up hydrogen equilibrium:

$$C_8H_{10}+H_2 <=> C_7H_8+CH_4 \qquad r_1=k_1\cdot[C_8H_{10}][H_2]^{1/2} \tag{1}$$

$$C_7H_8 + H_2 <=> C_6H_6+CH_4 \qquad r_2=k_2\cdot[C_7H_8][H_2]^{1/2} \tag{2}$$

where k_1=9.2 × 10^9exp(-220000/RT) and k_2= 4 × 10^9exp(-224000/RT). These reactions (1-2) have been used to simulate the gas phase decomposition of xylene and toluene in CVD reactors by Endo et al. [56], Kuwana et al. [57] and Wasel et al. [58]. They obtained an excellent agreement between the simulations and the experiment results at temperatures below 725°C. Kuwana et al. [59] conducted a xylene pyrolysis simulation using a detailed xylene oxidation reaction mechanism (160 species, 1175 reactions) constructed by Dagaut et al. [60,61] This comprehensive mechanism included the direct cleavage of R-CH_3 bonds, the generation of ethane, ethylene and acetylene, the decomposition of benzene, etc. Their simulation results agreed very well with the ones based on the global mechanism when the temperature was below 700°C. Meanwhile, the simulation showed that toluene, benzene and methane were the three major pyrolysis products from xylene.

Pyrolysis of Acetylene

Thermal pyrolysis of acetylene has been widely studied. Becker and Hüttinger [62] pointed out that the pyrolysis of acetylene at 1000°C generated three main gaseous reaction products-ethylene, benzene and methane. Back et al. [63] indicated that vinyl acetylene (C_4H_4) was one initial product from acetylene pyrolysis at the temperatures below 930°C. Kiefer et al. [64] also indicated that molecular polymerization was the main feature of homogenous acetylene pyrolysis at temperatures below 825°C. Based on a detailed reaction mechanism, Norinaga and Deutschmann [65] have recently modeled the pyrolysis of acetylene under one condition relevant to the CVD of pyrolytic carbon at 900°C. Their simulation predicted that acetylene was primarily transformed into vinylacetylene (86%), diacetylene (17 %), and benzene (7 %). The sensitivity analysis indicated the following three acetylene-related reactions were most sensitive among the hundreds:

$$2C_2H_2 <=> C_4H_4 \tag{3}$$

$$2C_2H_2 <=> C_4H_2+ H_2 \tag{4}$$

$$C_2H_2 + C_4H_4 <=> C_6H_6 \tag{5}$$

Based on the above discussion, the reactions 3-5 are chosen to describe the principal features of the acetylene pyrolysis in our CVD system. Given that the pyrolysis of acetylene takes place under the hydrogen atmosphere, the following reaction between H_2 and C_2H_2 is also taken into account,

$$C_2H_2 + H_2 <=> C_2H_4 \quad (6)$$

Table 1 summarizes the whole global mechanism used for the simulation of the decomposition of xylene, acetylene and ferrocene in our CVD reactor. And the reaction rate constant k is expressed using the modified Arrhenius equation,

$$k = AT^n \exp(-E_a/R_gT) \quad (7)$$

where A denotes the pre-exponential factor, T the temperature (K), n the exponent for temperature dependence of A; E_a activation energy (J mol^{-1}), and R_g the gas constant, 8.314 J mol^{-1} K^{-1}. The unit of pre-exponential factor is $(m^3\ mol^{-1})^{a-1}s^{-1}$, where a is the order of the reaction.

Modeling of Gas Phase Reaction Kinetics

By taking into account the physical and chemical phenomena inside the reactor, the chemical kinetics in the gas phase was numerically analyzed using COMSOL Multiphysics [41].

Figure 10 shows the comparison between the MS measurements and the simulation of the xylene and acetylene concentrations in the exhaust gas at different temperatures. The concentration variations were obtained from the mass spectra of the exhaust gas measured at the end of the reactor. A similar variation tendency was observed for both xylene and acetylene with the temperature increase. The increasing differences at higher temperatures could be mainly due to the fact that the simulation did not take into account the carbon decomposition reactions occurring on the substrate surface. The surface reactions at high temperatures resulted in the enlargement of the difference between the experiment and simulation results. However, the simulation can generally describe the consumptions of carbon and catalyst precursors under our experimental conditions. Thus, it can be used to generally predict the variation tendency of the chemical species.

Table 1. Global gas-phase reactions of xylene, acetylene and ferrocene in the CVD reactor and the Arrhenius parameters (in units of s^{-1}, m^3, mol^{-1} and J) [41]

Reaction	A	n	E (J mol^{-1})	Ref.
$C_8H_{10}+H_2 => C_7H_8+CH_4$	9.2×10^9	0	220000	Benson et al. [66]
$C_7H_8 + H_2 => C_6H_6+CH_4$	4×10^9	0	224000	Benson et al. [66]
$2C_2H_2 => C_4H_4$	5.5×10^6	0	154650	Norinaga et al. [65]
$C_4H_4 => 2C_2H_2$	8.65×10^{13}	0	312850	
$2C_2H_2 => C_4H_2+ H_2$	1.5×10^7	0	178670	Norinaga et al. [65]
$C_2H_2 + C_4H_4 => C_6H_6$	4.47×10^5	0	126000	Norinaga et al. [65]
$C_2H_2+ H_2 => C_2H_4$	1.41×10^{35}	-9.06	214000	Norinaga et al. [65]
$Fe(C_5H_5)_2+H_2 => Fe +2C_5H_6$	1.6×10^{11}	0	218000	Kuwana et al. [67]

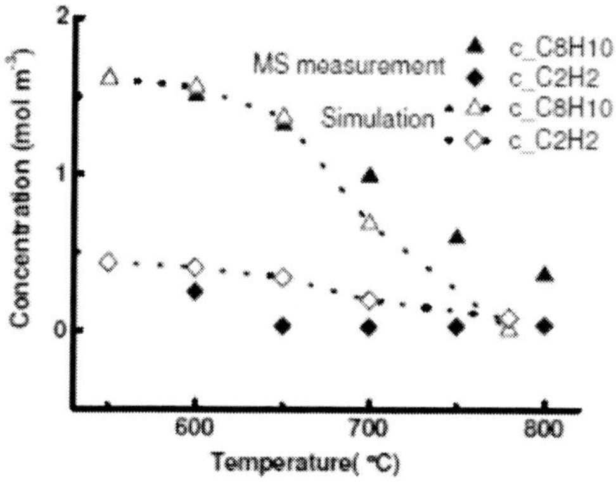

Figure 10. MS measurement and numerical simulation of the xylene and acetylene concentrations at the end of the heated zone of the CVD reactor at different temperatures.

As we know, the CNT growth on μAl_2O_3 follows the "bottom-up" mechanism [37]. The carbon deposition mainly occurs on the active catalyst particles located on the surface of alumina microparticles. Convection and diffusion are the two mass transfer mechanisms responsible for the feeding of carbon and iron species to sustain the continuous growth of CNTs. It is found that the gas convection is dominant in the center region of the reactor, while the mass diffusion does in the radial directions, especially in the layer near the substrate surface. These can be easily explained by the specific gas flow velocity and temperature field. The reacting gas has the highest flow rate in the center region, while lower near the tube inner wall and quartz plate due to the viscosity interaction. Also, the reactor has an intrinsic temperature gradient at each given temperature, especially in the inlet and outlet regions. This will generate variations of the chemical reaction rates and the relative concentration ratios between the species.

Figure 11a and b shows the temperature-dependent distribution of the chemical species concentrations along the centerline (tube axis) and a near-surface line (5 mm above quartz plate), respectively. Both xylene and ferrocene have very low decomposition rates at 550°C. Little of them is decomposed at the entrance stage. It is at ~0.2 m where the concentrations of C_7H_8, C_5H_6 and CH_4 reach up to 0.001 mol.m^{-3}. From there, these species concentrations gradually increase, and reach the maximum value of ~0.01 mol.m^{-3} near the outlet. This result corresponds very well with our experimental observations that the CNTs grow on μAl_2O_3 at a slow rate, and that the growth is only observed on the microparticles located in the zone after 0.2 m. The decomposition of xylene is largely accelerated above 700°C. The increased toluene decomposition generates large amount of benzene at the end of the reactor. The dominant species in the exhaust gas are toluene and methane at 700°C and benzene and methane at 780°C. By comparison, the species concentrations near the quartz plate surface are given in Figure 11b. The concentration variation tendency with temperature is found similar to the one along the centerline. However, due to the slow gas flow rates, both ferrocene and hydrocarbons have higher decomposition rates near the quartz plate than in the reactor center. In consequence, the product species such as methane, benzene, toluene and iron have relatively higher concentrations near the substrate surface, which favors the catalyst particle nucleation and CNT growth.

It is found that acetylene has high chemical activity which favors an efficient carbon feeding to CNT growth. Therefore, the addition of acetylene in the system provides more effective carbon source. An enhanced CNT growth is demonstrated at relatively low temperatures. According to our experimental observations, the addition of acetylene makes the CNT growth onset draw back about 10-15 cm on the quartz plate, compared with that without acetylene. This also indicates that a very small decomposition ratio of ferrocene could also generate CNT growth in the presence of acetylene. As shown in Figure 11 a and b, the pyrolysis of acetylene produces low concentration C_4H_4 in the rear zone of the reactor at 550°C. Besides C_4H_4, other pyrolysis species such as C_4H_2 and C_6H_6 also exhibit an increase in their concentrations when the temperature is higher than 600°C. Especially, their concentrations in the front zone of the reactor considerably increase at higher temperatures. Their concentration distributions in the reactor are displayed in Figure 11c. Obviously, the chemical reactions are non-isothermal process, and the concentration distribution of each species is space-dependent at the temperatures studied.

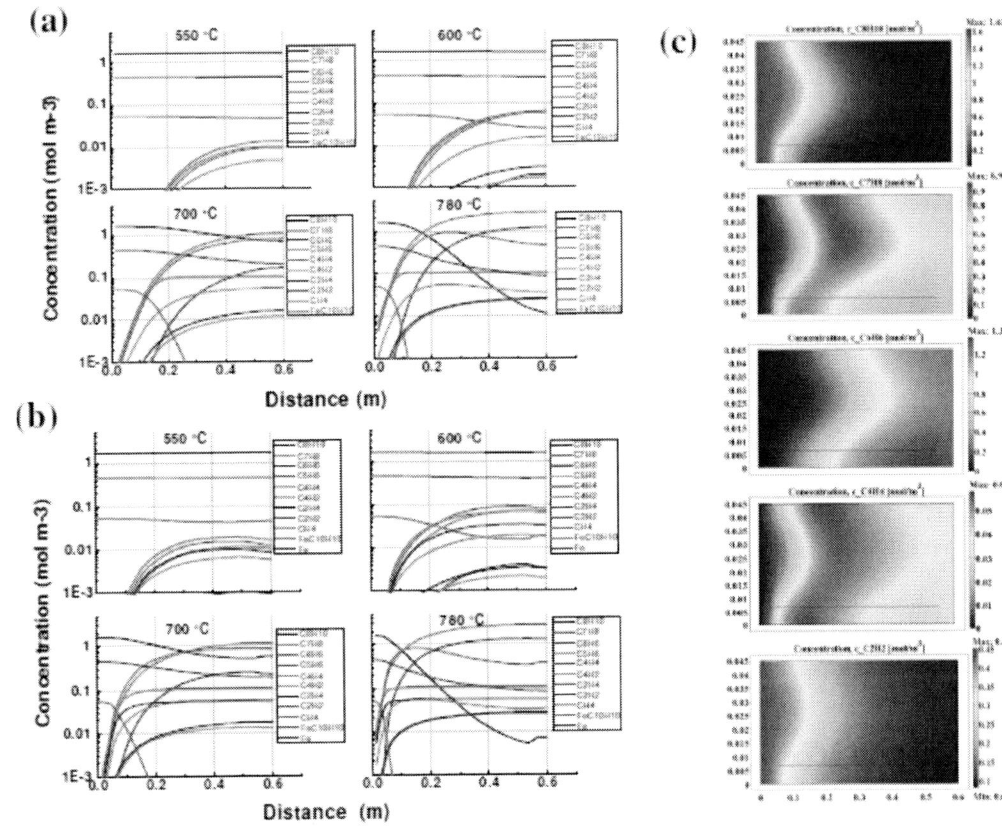

Figure 11. Chemical species concentration distributions obtained by simulating the decomposition of the xylene/acetylene mixture at different temperatures: (a) along the tube axis; (b) near the substrate surface (5 mm above the quartz plate), (c) Surface graph of the distribution of the main chemical species (in order of C_8H_{10}, C_7H_8, C_6H_6, C_4H_4, and C_2H_2 ~~and Fe~~) concentrations at 780°C [41].

Chemical Reaction Kinetics at Relatively High Temperatures

From the analysis of the reduced mechanism, the main products in the CVD process can be quantitatively identified with the MS measurements. However, the generation and variation of intermediate molecules and polycyclic aromatic hydrocarbons can not be described by this mechanism. Moreover, acetylene is known to play an important role both in CNT growth process and in soot formation process. This effect results from the interaction between different carbon sources used in the CVD process, which can be systematically analyzed by the detailed mechanism. Detailed mechanisms have also been proposed by many researchers (Errede and DeMaria [68]; Emdee et al. [69]; Richter et al. [70]; etc.) to better understand the decomposition process. Moreover, Gail and Dagaut [71] developed a detailed chemical kinetic reaction mechanism (160 species and 1175 reactions) to analyze the oxidation process of xylene. To investigate the non-isothermal structure in the reaction tube, the detailed mechanism to describe the decomposition process of xylene and acetylene was considered [42]. These researches involve various hydrocarbons as well as oxygen containing species and reactions. We construct the mechanism by collecting only hydrocarbon species and related chemical reactions. By choosing detailed mechanism with 88 species and 322 reactions [42], modeling of the decomposition process of xylene, acetylene and their mixture is realized. By modeling the detailed chemical reactions and related kinetic constants, the decomposition process of xylene, acetylene and their mixture at 800°C is presented in Figure 12-14. The experimental parameters used in this model are the same as that in our experimental conditions used [42].

In Figure 12, besides the main products (CH_4, C_6H_6, C_7H_8 and C_8H_{10}) described by the reduced mechanism, other species involving C_2H_2, C_2H_4, C_2H_6 and styrene are presented. These species are generated from the reaction between the pyrolysis products of xylene:

$$C_8H_{10}=CH_3C_6H_4CH_2+H \tag{8}$$

$$C_8H_{10}+H=C_7H_8+CH_3 \tag{9}$$

$$C_7H_8+H=C_6H_6+CH_3 \tag{10}$$

$$CH_3C_6H_4CH_2=BC_6H_7+C_2H_2 \tag{11}$$

$$CH_3+CH_3=C_2H_4+H_2 \tag{12}$$

$$CH_3+CH_3=C_2H_6 \tag{13}$$

$$C_6H_5C_2H_3=C_6H_6+C_2H_2 \tag{14}$$

In this process, most of the hydrocarbons are intermediate products, thus the final mole fractions are lower than 10^{-4}, almost invisible in the mass spectrum. As shown in Figure 15, the mass spectra are taken during the decomposition process of xylene, acetylene and their mixture, which can be correlated with the modelling results in Figure 12-14. When acetylene is introduced into the reaction system, the decomposition process of xylene is accelerated. As shown in Figure 13a, at 52 cm, xylene is decomposed completely. Comparing the result in

Figure 12a, at the end of the reaction tube, there is still some amount of xylene has not been decomposed. Moreover, when acetylene is used, the final mole fraction of CH_4, C_6H_6 and C_7H_8 decreases from 0.065, 0.041 and 0.017 (Figure 12a) to 0.035, 0.010 and 0.014 (Figure 13a), respectively. The information in Figure 15a-b also supports this result. Interestingly, with the addition of acetylene, various new species appear in the final products, involving C_5H_6 (cyclopentadiene), C_8H_6 (phenylethyne), C_8H_8 (styrene), C_8H_{10} (phenylethane), C_9H_8 (indene), $C_{10}H_8$ (naphthalene), $C_{11}H_{10}$ (methylnaphthalene). The related chemical reactions are:

$$C_5H_6+H=C_3H_3+C_2H_4 \quad (15)$$

$$CH_3C_6H_4CH_2=C_5H_6+C_3H_3 \quad (16)$$

$$C_6H_5+C_2H=C_6H_5C_2H \quad (17)$$

$$C_6H_5+C_2H_5=C_6H_5C_2H_5 \quad (18)$$

$$C_6H_5C_2H_5=C_6H_6+C_2H_4 \quad (19)$$

$$C_6H_4CH_3+C_2H_2=C_9H_8+H \quad (20)$$

$$C_6H_4C_2H+C_2H_2=C10H7 \quad (21)$$

$$C_{10}H_7+H=C_{10}H_8 \quad (22)$$

$$C_{10}H_7+CH_3=C_{10}H_7CH_3 \quad (23)$$

The presence of these species can be further confirmed by the mass spectrum in Figure 15b. Especially, the peak for C_9H_8 (116 amu) and $C_{10}H_8$ (128 amu) can be clearly identified in the graph at the top right corner in Figure 15b. It is worth to mention that these higher hydrocarbons also participate in the synthesis of CNTs. For instance, CNTs can be generated from the pyrolysis of naphthalene at the experimental temperature [72]. However, since their mole frictions are very low (approximately 10^{-3}), the contribution of the hydrocarbons to the growth of CNTs are negligible in our experiments. Figure 14 shows the decomposition process of acetylene. The main hydrocarbon products of this process are C_2H_4, C_4H_6 and C_6H_6, generated through the following reactions.

$$C_2H_3+C_2H=C_2H_2+C_2H_2 \quad (24)$$

$$C_2H_3+H=C_2H_2+H_2 \quad (25)$$

$$C_2H_4+H=C_2H_3+H_2 \quad (26)$$

$$C_2H_3+H=C_2H_4 \quad (27)$$

$$C_2H_3+C_2H_4=C_4H_6+H \quad (28)$$

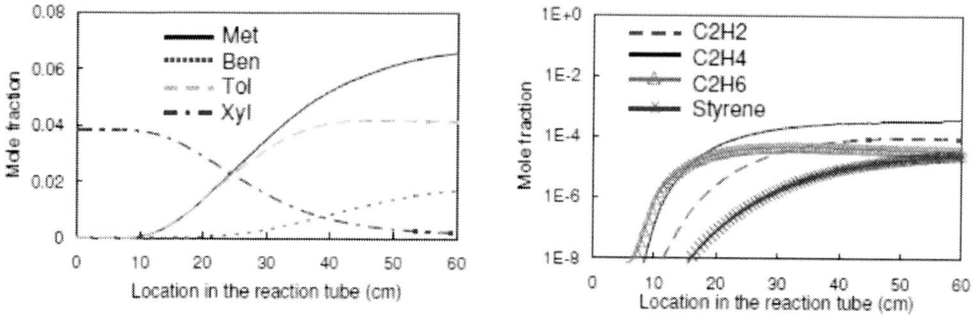

Figure 12. Variation of the different hydrocarbon species along the reaction tube in xylene decomposition process [42].

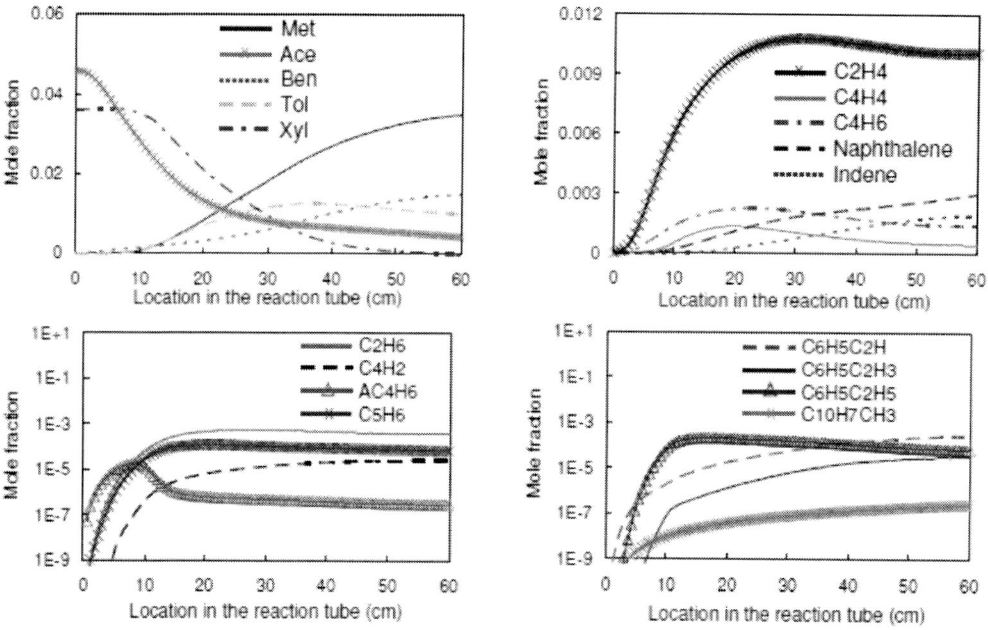

Figure 13. Variation of the different hydrocarbon species along the reaction tube in the decomposition process of acetylene/xylene mixture (acetylene flow rate: 50 ml min^{-1}) [42].

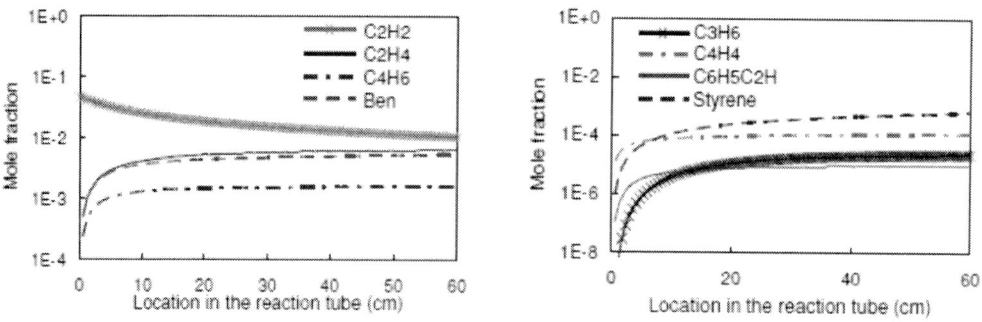

Figure 14. Variation of the different hydrocarbon species along the reaction tube in acetylene decomposition process [42].

$$C_4H_4=C_2H_2+C_2H_2 \quad (29)$$

$$C_4H_6=C_2H_4+C_2H_2 \quad (30)$$

$$C_4H_4+H=NC_4H_3+H_2 \quad (31)$$

$$NC_4H_3+C_2H_2=C_6H_5 \quad (32)$$

$$C_6H_6+H=C_6H_5+H_2 \quad (33)$$

NC_4H_3 represents 1-butene-3-yne-1-yl. The generation of benzene in the decomposition process of acetylene is active and can be confirmed by mass spectrum in Figure 15c. The high efficiency of acetylene in the production of CNTs discussed previously can be correlated with the decomposition process and species. As the main products in the gas phase, benzene and ethylene are reasonably considered as highly efficient carbon feedstock.

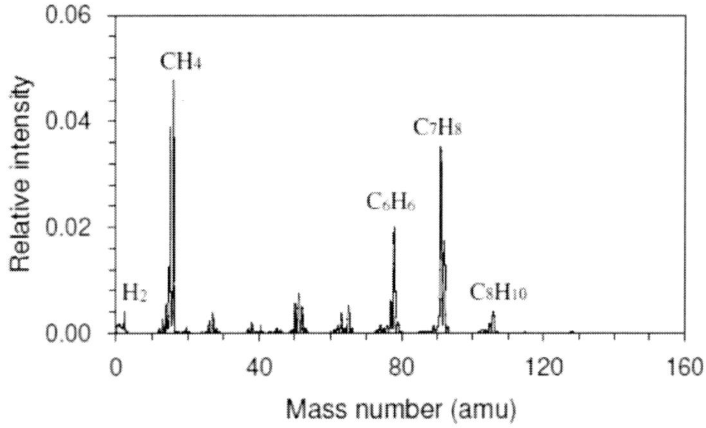

(a)

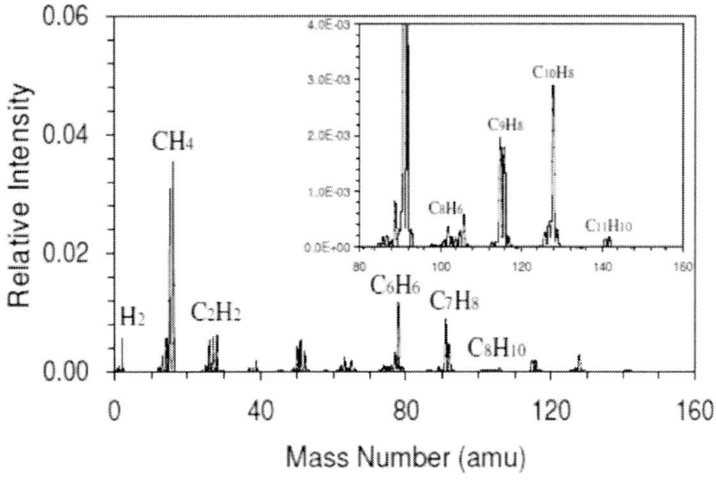

(b)

Figure 15. (Continued)

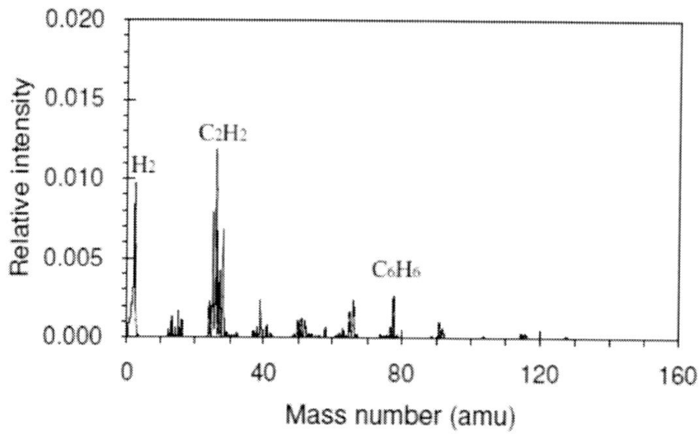

(c)

Figure 15. Mass spectra of gaseous components generated from the decomposition of: (a) xylene with flow rate 0.2 ml min^{-1}, (b) mixture of xylene (flow rate 0.2 ml min^{-1}) and acetylene (flow rate 50 ml min^{-1}), (c) acetylene with flow rate 50 ml min^{-1} [42].

Conclusion

In conclusion, xylene is a widely used aromatic hydrocarbon. Its dissociation was numerically and experimentally studied for the CVD synthesis of CNTs. Xylene undergoes a complex decomposition process which strongly depends on the reaction temperature. At relatively low temperatures (between 450 and 750°C), the direct pyrolysis of xylene and toluene into toluene and benzene are negligible, respectively. Instead, the hydrogen addition-exchange reactions being much easier, hydrogen can be used to catalyse the xylene dissociation at relatively low temperatures. The use of acetylene as an additional carbon source shows a synergistic effect as well. Hence, the decomposition of xylene is intensively promoted by the addition of acetylene. While acetylene can be polymerized to benzene, it will preferably react with toluene and benzene to form a series of various byproducts, hence decreasing the benzene concentration and promoting the decomposition of xylene. Furthermore, the xylene decomposition mechanisms were found to significantly influence the CNT growth, inducing strong variations on the diameter, length, and area number density which yields different CNT organization on micro-spherical substrates, namely ''short-dense-homogeneous'', ''six-branch'' and ''urchin-like''. Ultimately, such reported multiform hybrid structures are suitable for continuous and large-scale production. They are therefore promising to contribute to the enhancement of advanced multifunctional composite materials thanks to well-organized CNT dispersion and arrangement in a matrix.

References

[1] Lyons, TW; Guironnet, D; Findlater, M; Brookhart, M. Synthesis of p-xylene from ethylene. *J. Am. Chem. Soc.* 134, (2012), 15708-11.

[2] Williams, CL; Chang, C-C; Do, P; Nikbin, N; Caratzoulas, S; Vlachos, DG; et al. Cycloaddition of biomass-derived furans for catalytic production of renewable p-xylene. *ACS Catal*. 2, (2012), 935-9.
[3] Fan, Z; Yan, J; Wei, T; Zhi, L; Ning, G; Li, T; et al. Asymmetric supercapacitors based on graphene/MnO2 and activated carbon nanofiber electrodes with high power and energy density. *Adv. Funct. Mater*. 21, (2011), 2366-75.
[4] Li, W; Yuan, J; Dichiara, A; Lin, Y; Bai, J. The use of vertically aligned carbon nanotubes grown on SiC for in situ sensing of elastic and plastic deformation in electrically percolative epoxy composites. *Carbon*, 9, (2012), 4298-4301.
[5] Salvetat-Delmotte, J-P ; Rubio, A. Mechanical properties of carbon nanotubes: a fiber digest for beginners. *Carbon*. 40, (2002), 1729-34.
[6] Ruoff, RS; Lorents, DC. Mechanical and thermal-properties of carbon nanotubes. *Carbon*. 33, (1995), 925-30.
[7] Salvetat, JP; Bonard, JM; Thomson, NH; Kulik, AJ; Forro, L; Benoit, W; et al. Mechanical properties of carbon nanotubes. *Appl. Phys. A-Mater. Sci. Proc*. 69, (1999), 255-60.
[8] Kim, P; Shi, L; Majumdar, A; McEuen, PL. Thermal Transport Measurements of Individual Multiwalled Nanotubes. *Phys. Rev. Lett*. 87, (2001), 215502.
[9] Bozlar, M; He, D; Bai, J; Chalopin, Y; Mingo, N; Volz, S. Carbon nanotube microarchitectures for enhanced thermal conduction at ultra low mass fraction in polymer composites. *Adv. Mater*. 22, (2010), 1654.
[10] Ajayan, PM. Nanotubes from carbon. *Chem. Rev*. 99, (1999), 1787-800.
[11] Ebbesen, TW; Lezec, HJ; Hiura, H; Bennett, JW; Ghaemi, HF; Thio, T. Electrical conductivity of individual carbon nanotubes. *Nature*. 382, (1996), 54-56.
[12] Wilder, JWG; Venema, LC; Rinzler, AG; Smalley, RE; Dekker, C. Electronic structure of atomically resolved carbon nanotubes. *Nature*. 391, (1998), 59-62.
[13] Iijima, S. Helical microtubules of graphitic carbon. *Nature*. 354, (1991), 56-58.
[14] Iijima, S; Ichihashi, T. Single-shell carbon nanotubes of 1-nm diameter. *Nature*. 363, (1993), 603-605.
[15] Ebbesen, TW; Ajayan, PM. Large-scale synthesis of carbon nanotubes. *Nature*. 358, (1992), 220-222.
[16] He, DL; Zhao, TK; Liu, YN; Zhu, JW; Yu, G; Ge, LL. The effect of electric current on the synthesis of single-walled carbon nanotubes by temperature controlled arc discharge. *Dia. Rel. Mater*. 16, (2007), 1722-6.
[17] He, DL; Liu, YN; Zhao, TK; Zhu, JW; Yu, G. Effect of metal oxide and oxygen on the growth of single-walled carbon nanotubes by electric arc discharge. *J. Nanopart. Res*. 10, (2008), 409-414.
[18] Thess, A; Lee, R; Nikolaev, P; Dai, HJ; Petit, P; Robert, J; et al. Crystalline ropes of metallic carbon nanotubes. *Science*. 273, (1996), 483-7.
[19] Joseyacaman, M; Mikiyoshida, M; Rendon, L; Santiesteban, JG. Catalytic growth of carbon microtubules with fullerene structure. *Appl. Phys. Lett*. 62, (1993), 202-4.
[20] Dai, HJ. Carbon nanotubes: opportunities and challenges. *Surf. Sci*. 500, (2002), 218-41.
[21] Dai, HJ. Carbon nanotubes: Synthesis, integration, and properties. *Accounts Chem. Res*. 35, (2002), 1035-44.

[22] Ci, LJ; Bai, J. Novel micro/nanoscale hybrid reinforcement: Multiwalled carbon nanotubes on SiC particles. *Adv. Mater*. 16, (2004), 2021.
[23] Chen, Y; Conway, MJ; FitzGerald, JD; Williams, JS; Chadderton, LT. The nucleation and growth of carbon nanotubes in a mechano-thermal process. *Carbon*. 42, (2004), 1543-8.
[24] Vander Wal, RL; Ticich, TM; Curtis, VE. Diffusion flame synthesis of single-walled carbon nanotubes. *Chem. Phys. Lett*. 323, (2000), 217-23.
[25] Yuan, L; Saito, K; Pan, C; Williams, FA; Gordon, AS. Nanotubes from methane flames. *Chem. Phys. Lett*. 340, (2001), 237-41.
[26] Yuan, LM; Saito, K; Hu, WC; Chen, Z. Ethylene flame synthesis of well-aligned multi-walled carbon nanotubes. *Chem. Phys. Lett*. 346, (2001), 23-8.
[27] Matveev, AT; Golberg, D; Novikov, VP; Klimkovich, LL; Bando, Y. Synthesis of carbon nanotubes below room temperature. *Carbon*. 39, (2001), 155-8.
[28] Hsu, WK; Terrones, M; Hare, JP; Terrones, H; Kroto, HW; Walton, DRM. Electrolytic formation of carbon nanostructures. *Chem. Phys. Lett*. 262, (1996), 161-6.
[29] Hsu, WK; Hare, JP; Terrones, M; Kroto, HW; Walton, DRM; Harris, PJF. Condensed-phase nanotubes. *Nature*. 377, (1995), 687-9.
[30] Bai, J; Hamon, AL; Marraud, A; Jouffrey, B; Zymla, V. Synthesis of SWNTs and MWNTs by a molten salt, (NaCl), method. *Chem. Phys. Lett*. 365, (2002), 184-8.
[31] Laplaze, D; Bernier, P; Maser, WK; Flamant, G; Guillard, T; Loiseau, A. Carbon nanotubes: The solar approach. *Carbon*. 36, (1998), 685-8.
[32] Cho, W-S, Hamada, E; Kondo, Y; Takayanagi, K. Synthesis of carbon nanotubes from bulk polymer. *Appl. Phys. Lett*. 69, (1996), 278-9.
[33] Li, YL; Yu, YD; Liang, Y. A novel method for synthesis of carbon nanotubes: Low temperature solid pyrolysis. *J. Mater. Res*. 12, (1997), 1678-80.
[34] He, D; Bai, J. Acetylene-Enhanced Growth of Carbon Nanotubes on Ceramic Microparticles for Multi-Scale Hybrid Structures. *Chem. Vap. Dep*. 17, (2011), 98-106.
[35] He, D; Bozlar, M; Genestoux, M; Bai, J. Diameter- and length-dependent self-organizations of multi-walled carbon nanotubes on spherical alumina microparticles. *Carbon*. 48, (2010), 1159-70.
[36] He, D; Li, H; Li, W; Haghi-Ashtiani, P; Lejay, P; Bai, J. Growth of carbon nanotubes in six orthogonal directions on spherical alumina microparticles. *Carbon*. 49, (2011), 2273-86.
[37] Dichiara, A; Bai, J. The growth of carbon nanotube multilayers on ceramic u-particles by catalytic chemical vapour deposition. *Dia. Rel. Mater*. 29, (2012), 52-8.
[38] Dichiara, A; Yuan, J; Yao, S-H; Sylvestre, A; Bai, J. Chemical Vapor Deposition Synthesis of Carbon Nanotube-Graphene Nanosheet Hybrids and Their Application in Polymer Composites. *J. Nanosci. Nanotechnol*. 12, (2012), 6935-40.
[39] de Villoria, RG; Hart, AJ; Wardle, BL. Continuous High-Yield Production of vertically aligned carbon nanotubes on 2D and 3D substrates. *ACS Nano*. 5, (2011), 4850-7.
[40] Kim, DY; Sugime, H; Hasegawa, K; Osawa, T; Noda, S. Sub-millimeter-long carbon nanotubes repeatedly grown on and separated from ceramic beads in a single fluidized bed reactor. *Carbon*. 49, (2011), 1972-9.
[41] He, D; Li, H; Bai, J. Experimental and numerical investigation of the position-dependent growth of carbon nanotube-alumina microparticle hybrid structures in a horizontal CVD reactor. *Carbon*. 49, (2011), 5359-72.

[42] Li, H; He, D; Li, T; Genestoux, M; Bai, J. Chemical kinetics of catalytic chemical vapor deposition of an acetylene/xylene mixture for improved carbon nanotube production. *Carbon*. 48, (2010), 4330-42.
[43] He, D. Formation of hybrid structures of carbon nanotubes and alumina microparticles by CVD method: Mechanisms and Chemical kinetics. *Chatenay-Malabry*, (Fr),: Ecole Centrale Paris; 2010.
[44] Eres, G; Puretzky, AA; Geohegan, DB; Cui, H. *In situ* control of the catalyst efficiency in chemical vapor deposition of vertically aligned carbon nanotubes on predeposited metal catalyst films. *Appl. Phys. Lett.* 84, (2004), 1759-61.
[45] Benson, SW; Shaw, R. Kinetics and mechanism of hydrogenolyses. The addition of hydrogen atoms to propylene, toluene, and xylene. *J. Chem. Phys.* 47, (1967), 4052-5.
[46] Yang, KL; Yang, RT. The accelerating and retarding effects of hydrogen on carbon deposition on metal-surfaces. *Carbon*. 24, (1986), 687-93.
[47] Behr, MJ; Gaulding, EA; Mkhoyan, KA; Aydil, ES. Hydrogen etching and cutting of multiwall carbon nanotubes. *J. Vac. Sci. & Technol. B*. 28, (2010), 1187-94.
[48] Borjesson, A; Bolton, K. First Principles Studies of the Effect of Ostwald Ripening on Carbon Nanotube Chirality Distributions. *ACS Nano*. 5, (2011), 771-9.
[49] Di Vece, M; Grandjean, D; Van Bael, MJ; Romero, CP; Wang, X; Decoster, S; et al. Hydrogen-induced ostwald ripening at room temperature in a Pd nanocluster film. *Phys. Rev. Lett.* 13, (2008), 100-23.
[50] Nolan, PE; Schabel, MJ; Lynch, DC; Cutler, AH. Hydrogen control of carbon deposit morphology. *Carbon*. 33, (1995), 79-85.
[51] Lin, Y; Dichiara, A; He, D; Haghi-Ashtiani, P; Bai; J. Uniform diameter multi-walled carbon nanotubes with a controlled wall number obtained by a simple chemical vapor deposition method. *Chem. Phys. Lett.* 554, (2012), 142.
[52] Jansen, EF; Wallace, JM. Formation of Benzene and Toluene from Acetylene-14C in the Avocado. *J. Biol. Chem.* 240, (1965), 1042-4.
[53] Jungwirthova, I; Kessmodel, LL. Benzene formation from acetylene on Pd(1 1 1),: a high-resolution electron energy loss spectroscopy study. Kidlington, UK: Elsevier 2001.
[54] Schmitt, TC; Bins, AS; Miller, DW; Bins, AR; Lupu D, D; Trigwell, S; et al. Analysis of effluent gases during the CCVD growth of multi-wall carbon nanotubes from acetylene. *Carbon*, 44, (2006), 2032-8.
[55] Rodgers, AS; Golden, DM; Benson, SW. Kinetics of reaction of iodobenzene and hydrogen iodide: Heat of formation of phenyl radical and its implications on reactivity of benzene. *J. Am. Chem. Soc.* 89, (1967), 4578-83.
[56] Endo, H; Kuwana, K; Saito, K; Qian, D; Andrews, R; Grulke, EA. CFD prediction of carbon nanotube production rate in a CVD reactor. *Chem. Phys. Lett.* 387, (2004), 307-11.
[57] Kuwana, K; Endo, H; Saito, K; Qian, D; Andrews, R; Grulke, EA. Catalyst deactivation in CVD synthesis of carbon nanotubes. *Carbon*. 43, (2005), 253-60.
[58] Wasel, W; Kuwana, K; Saito, K. Chemical and thermal structures of a xylene-based CVD reactor to synthesize carbon nanotubes. *Chem. Phys. Lett.* 422, (2006), 470-4.
[59] Kuwana, K; Li, T; Saito, K. Gas-phase reactions during CVD synthesis of carbon nanotubes: Insights via numerical experiments. *Chem. Eng. Sci.* 61, (2006), 6718-26.

[60] Dagaut, P; Pengloan, G; Ristori, A. Oxidation, ignition and combustion of toluene: Experimental and detailed chemical kinetic modeling. *Phys. Chem. Chem. Phys.* 4, (2002), 1846-54.

[61] Gail, S; Dagaut, P. Experimental kinetic study of the oxidation of p-xylene in a JSR and comprehensive detailed chemical kinetic modeling. *Combustion and Flame.* 141, (2005), 281-97.

[62] Becker, A; Huttinger, KJ. Chemistry and kinetics of chemical vapor deposition of pyrocarbon - II - Pyrocarbon deposition from ethylene, acetylene and 1,3-butadiene in the low temperature regime. *Carbon.* 36, (1998), 177-99.

[63] Back, MH. Mechanism of pyrolysis of acetylene. *Canadian J. Chem.* 49, (1971), 2199.

[64] Kiefer, JH; Vondrasek, WA. The mechanism of the homogeneous pyrolysis of acetylene. *Int. J. Chem. Kinet.* 22, (1990), 747-86.

[65] Norinaga, K; Deutschmann, O. Detailed Kinetic Modeling of Gas-Phase Reactions in the Chemical Vapor Deposition of Carbon from Light Hydrocarbons. *Ind. Eng. Chem. Res.* 46, (2007), 3547-57.

[66] Benson, SW; Shaw, R. Kinetics and mechanism of hydrogenolyses: Addition of hydrogen atoms to propylene toluene and xylene. *J. Chem. Phys.* 47, (1967), 4052.

[67] Kuwana, K; Saito, K. Modeling ferrocene reactions and iron nanoparticle formation: Application to CVD synthesis of carbon nanotubes. *Proc. Comb. Inst.* 31, (2007), 1857-64.

[68] Errede, LA; DeMaria, F. The chemistry of xylenes. XV. The kinetics of fast flow pyrolysis of p-xylene. *J. Phys. Chem.* 66, (1962), 2664-72.

[69] Emdee, JL; Brezinsky, K; Glassman, I. High-temperature oxidation mechanisms of m- and p-xylene. *J. Phys. Chem.* 95, (1991), 1626-35.

[70] Richter, H; Howard, JB. Formation of polycyclic aromatic hydrocarbons and their growth to soot: a review of chemical reaction pathways. *Prog. E. Comb. Sci.* 26, (2000), 565-608.

[71] Gaïl, S; Dagaut, P. Experimental kinetic study of the oxidation of p-xylene in a JSR and comprehensive detailed chemical kinetic modeling. *Combustion and Flame.* 141, (2005), 281-97.

[72] Puengjinda, P; Sano, N; Tanthapanichakoon, W; Charinpanitkul, T. Selective synthesis of carbon nanotubes and nanocapsules using naphthalene pyrolysis assisted with ferrocene. *J. Ind. Eng. Chem.* 15, (2009), 375-80.

In: Xylenes
Editor: Michael Olawale Daramola
ISBN: 978-1-62808-342-2

Chapter 9

BIOLOGICAL INDICATORS OF EXPOSURE TO XYLENE AT WORK: METHODS AND LIMITATIONS

L. Caporossi and B. Papaleo
INAIL- Research Area. Department of Occupational Medicine,
Monteporzio Catone (RM), Italy

ABSTRACT

Over the years the occupational exposure to xylenes involves a considerable number of workers. The researchers defined analytical methods capable to identify the absorbed dose by the individual subject and the use of different indicators of exposure was proposed (blood xylene, methyl hippuric acids, ortho, meta and para, in urine, etc.). However the instrumental analytical methods were increasingly more refined with the use and dissemination of techniques associated with mass spectrometry and liquid and gas chromatography, as a consequence in many cases the limits proposed by national and international organisms (ACGIH, DFG, etc.), commonly based on UV detectors, are less indicative. In UV analysis possible confounding factors may be a lot, especially starting from complex matrices such as biological (blood, urine) and then requires a very careful treatment of the sample, often time-consuming, and a good chromatographic separation in order to ensure that the detection of the analyte of interest is correct. There are also new opportunities, such as the determination in various biological matrices, like saliva, or of new indicators in the determination of individual metabolites of the individual isomers, for a more precise determination of the exposure situations. We draw an overview of analytical methods applied for the realization of the biological monitoring of workers exposed to xylenes, with regard to existing biological indicators and the new proposals, showing limits for the prevention of health risks.

Keywords: Exposure, xylenes, indicators

Introduction

Xylene is used at industrial level in many production cycles mainly acting as solvent and diluent (production of paints and dyes, rubber industry, leather industry, activities of typography, production of various artifacts) or as a degreasing agent or as a component of fuel (in aviation) or as a reagent for the synthesis of other chemicals (petrochemicals and chemical industry, textile industry of synthetic fabrics, plastics production).

This wide use in many industrial contexts of production determines the possibility/ probability of exposure for workers assigned to specific tasks that involve handling xylene. Xylene vapor is absorbed rapidly from the lungs and xylene liquid and vapors are absorbed slowly through the skin.

The main effect of inhaling xylene vapor is depression of the central nervous system (CNS) with symptoms such as headache, dizziness, nausea and vomiting. Damage have been reported in cases of severe exposure to xylene with bone marrow hyperplasia, liver enlargement, liver necrosis, and nephrosis [1]. Volunteers have tolerated 100 ppm, but higher concentrations become objectionable. Xylene is an irritant of the eyes and mucous membranes at concentrations below 200 ppm and it is narcotic at high concentrations [2]. Irritation of the nose and throat can occur at approximately 200 ppm after 3 to 5 minutes. Exposures estimated at 700 ppm have caused nausea and vomiting. High concentrations (approximately 10,000 ppm) could cause incoordination, loss of consciousness, respiratory failure and death. In some cases a potentially fatal accumulation of fluid in the lungs (pulmonary edema) may result as observed in three workers exposed to xylene concentrations of approximately 10,000 ppm for 18.5 hours: one died and two recovered slowly after a period of unconsciousness and retrograde amnesia; disturbances of liver and kidney function were noted in these workers [2; 3]. Symptoms of pulmonary edema, such as shortness of breath and difficulty breathing, may be delayed several hours after exposure.

The estimated oral LD (50) (letal dose, which consists in the dose administered to a population of laboratory animals which determined the death of 50% of subjects) for humans is 50 mg/kg [4]. Ingestion of xylene causes gastrointestinal distress and may cause toxic hepatitis. A worker exposed to the vapors of a solvent containing 75 percent of xylene (approximate airborne xylene concentration of 60 to 350 ppm) developed giddiness, anorexia, and vomiting. Repeated contact of the skin with xylene causes drying and dermatitis [5].

The many variables which affect the absorption, metabolism and clearance of xylene include exercise, alcohol intake [6], cigarette smoking, co-exposure to other solvents, gender, gastrointestinal, hepatic and renal pathology. Concomitant exposure to xylene and other solvents [7], including toluene, affected hematological parameters, liver size, liver enzymes, auditory memory, visual abstraction and vibration threshold in the toes. Normal metabolic pathways were altered and significant increases in some serum bile acids may reflect early liver damage [8].

Biomarkers are becoming increasingly important in toxicology, many research groups are carrying out studies to develop biomarkers of exposure to chemicals and apply these to human monitoring. There is considerable interest in the use and application of biomarkers to identify the nature and amounts of chemical exposures in occupational and environmental situations [9].

The environmental monitoring is usually used to evaluate the workers exposure to chemicals but the association with biological monitoring provides better information on the amount of substance absorbed by the worker.

Biological Indicators for Biomonitoring of Workers Exposed to Xylene

The identification of indicators of internal dose requires knowledge of the toxicokinetic of xylene: information about the distribution of the chemical compound in the human organism as a result of exposure; metabolic reactions which may be encountered before the step of excretion. This compound is absorbed through inhalation, dermal and ingestion. When it has inhaled, at pulmonary level is presented a rapid absorption, for ingestion it can have rapid effects as a result of a rapid absorption, for dermal it can have an absorption but usually slow and in the case of prolonged contact over time; xylene undergoes a metabolic biotransformation that leads to the formation (for 95%) of methylhippuric acids (with isomerism corresponding to the starting isomer of the xylene) [10].

The o-xylene, for example, is oxidized to acid o-metyltoluic and after conjugation with glycine, to o-methylhippuric acid, the reaction gives rise also to 2,3-dimethylphenol and 3,4-dimethylphenol. The m-xylene forms m-toluic acid which is subsequently conjugate and it forms m-metylhippuric acid and similarly the p-xylene goes to synthesize the corresponding acid, p-toluic acid, which is then transformed into p-metylhippuric acid [10;11].

The elimination of xylene isomers is accomplished very slightly by the pulmonary route in exhaled air (on average 4.9%) [10].

Xylene in Exhaled Air

Some authors [12] have used the determination of xylenes in exhaled air as indicators of exposure to xylene, the study predicted the survey of 800 subjects, who were recruited between 1981 and 1987 in the United States, the readings showed levels correlated with environmental exposure conditions collected; certainly this kind of indicator presents many difficulties in application, not least the problems related to the complexity of the sampling phase, in order to obtain the representativeness of the data, as well as confounding factors easily detectable as for smokers, whose levels of xylenes have been also showed consistently higher than two-fold compared to non-smokers in the same conditions of exposure [13].

Methylhippuric Acids in Urine

The excretion of methylhippuric acids (MHAs) in the urine affects about 95% of xylene absorbed, this excretion occurs rather quickly with a first peak at 2-5 hours after exposure and a second peak excretion after an average of 30 hours [14]. For this reason urinary MHAs are biomarkers commonly used; typically was performed a comprehensive evaluation of the three isomers.

At regulatory level [15] they have been proposed for the environmental monitoring a Threshold Limit Value-Time Weighted Average- (TLV- TWA) expressed as the sum of the three xylene isomers (ppm) while, for the biological monitoring, the sum of the quantities of the three methylhippuric acids present in urine samples collected at the end of the working shift; historically to obtain a biologically useful indicators was suggested that the urinary concentration should be corrected for creatinine [16].

These compounds have proved useful as biomarkers as there are close correlations with environmental concentrations [17, 18, 19].

At the same exposure levels in those subjects with the habit of drinking alcohol, excretion of metabolites of xylenes was almost 20% lower than non-drinkers [20]: this trend was justified assuming a slowdown in metabolic process and then excretion. Another factor leading to a slow excretion is the body mass index, because was highlighted a certain capacity for accumulation in the lipid tissues that caused a slowed metabolism, due to a lower metabolic availability of xylene and then a slower excretion of its metabolites. Some studies [21] have investigated the incidence of sex in relation to the metabolic pathways in which m-xylene participates; to assess whether, in the case of inhalation, was highlight a different kinetics between men and women m-xylene was monitored in expired air, in blood, in saliva and urine by comparing the urinary m-methylhippuric acid, the survey conducted on a sample of 9 women and 8 men did not lead to confirm the hypothesis, there were no significant differences between the sexes in terms of kinetics of excretion.

For a correct assessment of the dose absorbed is necessary to know that metabolism of xylenes is significantly reduced in smokers and drinkers [22], taking aspirin reduces the excretion of m-methylhippuric acid [23] and a contemporary exposure during the working shift to m-xylene and methyl ethyl ketone or ethylbenzene modifies the kinetics of excretion of m-methylhippuric acid [24, 25, 26].

Xylene in Blood

As an alternative to the value of urinary metabolites of xylene the Deutsche Forschungsgemeinschaft (DFG) [27] proposed to conduct biological monitoring using blood concentrations of xylene. The limit value proposed by the DFG for this indicator amounted to 1.5g/L (in the blood). This hypothesis is based on studies conducted at industry level [14]: they showed statistically significant linear correlation between the environmental levels of xylenes and their determination in blood; also a correlation with MHAs in urine was identified. In surveys of exposed workers using xylenes in blood as biomarkers, it became evident as the smoking habit was not going to be an analytic confounding factor [28].

Xylene in Urine

Another proposed biomarker of exposure was the solvent (the three isomers: o- m-and p-xylene), in the urine, [29]. This proposal emerged from a study in which the analytical determination of urinary xylene was placed in correlation with the values of environmental concentrations and significant correlations were identified, however further investigation not confirmed this situation. The percentage of solvent that can be found in urine compared to metabolites is equivalent to just over 1% of the dose absorbed, for which the determination of the solvent unmetabolized in urine would require analytical methods considerably more sensitive than the determination of urinary metabolites.

Xylene in Saliva

Biological monitoring of occupational exposure to chemicals always requires the workers concerned to be willing to take part in the investigation, when a blood sample is needed, however workers are not always willing because the sampling is invasive. An analytical matrix that is easier to sample is the saliva, however salivary concentration of molecules of interest are normally much lower than plasma, so highly sensitive analytical equipment is essential to reach the low levels of detection, especially if workers are not exposed to high doses [30]. Some authors proposed the use of salivary xylene as biomarker, in one case a comparison was carried out among blood, urine, expired air and saliva levels [31]: results showed that data in saliva were parallel to that in blood but on average seven time lower; others [32] proposed the salivary xylene as biomarker but no correlation with other indicators of internal dose were indicated.

Analytical Methods for Biological Monitoring

Xylene in Blood

The use of blood (plasma or serum) as matrix to investigate the internal dose, for subjects exposed to xylene, is not a common analysis. The instrumental equipment necessary for the dosage in blood is usually the gas-chromatography/ mass spectrometry (GC/MS), with a pre-treatment of the sample using the solid phase micro extraction (SPME) [33], so the complexity of the matrix was eliminated and the analysis carried out with good accuracy and precision even at low doses [34].

Xylene in Urine

The application of the measurement of urinary excretion of unmodified solvent for the biological monitoring has been investigated in many papers [35,36]. The correlations observed were generally good: r^2 values range from 0.50-0.97, and the majority are between 0.84 and 0.90. The regression lines reported for the solvent in different studies present some variability: this is possibly due to an inadequate control of factors influencing the relationship between external dose and absorption, such as differences in body burden, work load, individual characteristics, etc. The results reported in the literature show that measuring of urinary excretion of unmodified solvent provides a highly sensitive and specific exposure index and can also be applied for low levels [37]. As said the percentage of xylene that is in urine unmetabolized is a 1% of the absorbed dose, for this reason are necessary instrumental methods of particular sensibility. A study on 65 workers showed a good correlation between xylene in blood and in urine and environmental monitoring [34] using a method with a SPME and GC/MS analysis. The same instrument was used by Tacheuchi et al, (35) in this study were collected urine samples from 86 workers: the exposure-excretion relationship was examined by multiple regression analysis, they found that the use of mother solvent in urine as biomarker is possible in cases of xylenes, either for three isomers in combination or separately. In this case further studies are needed for an adequate assessment of biological limit values, evaluating the reproducibility of regression lines, to investigate the aspects influencing the correlation between external dose and urinary excretion.

Xylene in Breath

The use of expired air for biological monitoring is not so common. Three breath sampling and analytical methods were tested following exposure of 12 subjects for 4 hr to 75 ppm m-xylene in a controlled environmental chamber. Expired breath for m-xylene was analyzed using gas chromatography with flame ionization detection (GC/FID). Comparison of the average and individual mixed to alveolar ratios of m-xylene and carbon dioxide showed that the sampling method is essential to obtain correct results; in this study two type of sampling were tested: only mainstream mixed sampling was accurate [38].

Results "on field"[39] indicated that breath concentrations of xylene were significantly correlated with environmental personal monitoring concentrations, even if the multiple regression analysis showed a coefficient of correlation not so high ($r^2 = 0.665$).

Metabolites of Xylene in Urine

Various methods have been published for the quantitative determination of MHAs [40,41, 42, 43, 44, 45].

The instrumental technique used in most cases provides the biological monitoring through the sampling of urine at the end of the work shift, without the use of preservatives or stabilizers, and a storage at -18° C for a maximum period of 6 months within which to execute the analysis [42]. The analytical determination is performed commonly with high performance liquid chromatography (HPLC) in most cases with UV detection (analytical wavelengths in the range from 207 to 270 nm) and isocratic elution [46,47,41,48,17,18]. To reduce the variability associated with sample preparation are generally used methods of internal standardization (with 3-hydroxybenzoic acid, 4-hydroxybenzoic acid, methylbenzo-alanine or p-chlorophenol).

The National Institute for Occupational Safety and Health (NIOSH) considered the use of HPLC equipment as the reference method (NIOSH 8301), however this method cannot separately quantitate each isomer of m- and p-MHA.

Published liquid chromatographic methods have the disadvantage that they do not allow the separation of the meta and para MHA isomers. This separation is in fact problematic and some researchers suggest adding tetra n-butylammonium bromide [49] or tetrahydrofurane [50] or beta-cyclodextrin [51, 20, 52, 53, 43] to the mobile phase; the latter compound offers advantages in terms of cost, separation capacity and ease of use, in fact this chemical easily forms inclusion complexes with a wide range of compounds, but especially with the aromatic ring [54, 55].

Some authors have proposed the use of capillary electrophoresis, which seems promising, allowing a reduction in the time of analysis with a satisfactory resolution of the isomers meta and para [56; 57]. With this analytical method were assayed MHAs together with urinary metabolites of toluene and styrene. The comparison of this type of analysis with HPLC shows a good correlation especially for subjects exposed to the vapors of a solvent mixture of toluene and xylene [56].

Recently in literature [58] a method using reversed-phase ultra-high performance liquid chromatography (UPLC) coupled with electrospray ionization tandem mass spectrometry (ESI/MS/MS) was published to simultaneously quantify 28 urinary volatile organic compounds metabolites, in 1203 subjects non smokers and 347 smokers, as biomarkers of exposure; the method was accurate (mean accuracy for spiked matrix ranged from 84-104%), sensitive (Limit of detection- LOD ranged from 0.5 to 20 ng/mL) and precise (RSD ranged from 2.5 to 11%).

Other authors have set gas chromatographic methods with the use of detectors FID or electron capture (ECD) or mass spectrometry (MS), using usually internal standards (tridecanoic acid, benzyl mercapturic acid...) [40,42,59,45,60]. Gas chromatography [61,62,63] is specific, with excellent precision and accuracy, but the sample has to be derivatized, e.g. using diazomethane [61], trimethyl-silyl derivatives [64] or methanol in hydrochloric acid [59], resulting in a longer sample work-up procedure and hazardous analytes as well as by products. In GC is usually present a pretreatment of the sample, consisting generally in a liquid/liquid extraction (with ethyl acetate) or in a solid phase extraction, with SPE cartridges C18, using HCl and heptane for the purification of the sample and a mixture chloroform/methanol 92:8 v/v for the extraction step [42]. Following this step the samples are esterified prior to analysis by GC.

Most of the methods of dosage by HPLC does not provide for a pretreatment of the sample, or purifications, but only a simple dilution in water or with the mobile phase (dilution factors from 1:10 to 1:100) [20,50,65]; other authors have, instead, extracted from urine with acidified organic solvent: after evaporation of the solvent and dissolving the residue in the mobile phase, the sample was ready for analysis by HPLC [40;53].

The application of chromatography seems particularly advantageous in terms of results. All the authors state recoveries greater than 90%, with a precision between 2-10%, expressed as coefficient of variation, to the different concentration levels tested. For the spectrophotometric determinations in UV was detected the presence of an interfering peak, for cases in which the urine sample is not treated in any way before analysis, very close to the peak of the o-MHA, this therefore requires a careful control of the working conditions and, where possible, the application of some variable (such as the use of β-cyclodextrin in the mobile phase) to prevent possible mistakes in interpretation of the chromatogram.

The methods in HPLC/UV are not optimal for the determination of urinary levels of MHAs if environmental concentration is equal to or less than 1mg/m^3. In this case is better to use detectors more sensitive and selective, as the mass spectrometer or the use of gas chromatography. [66].

In order to identify alternative indicators some authors have considered the possibility of analyzing minor metabolites of xylene in urine [67,68,65]. In one study was detected dimethylphenylmercapturic acid (DPMA) in urine [67]; the unequivocal detection of DPMA was carried out with LC/MS/MS, according to a rough calculation DPMA is only formed in a ratio of 0.0003% of absorbed xylene, DPMA in urine is not enough sensitive for general biomonitoring purposes due to the low ratio of excretion. S-benzyl methyl mercapturic acid (S-BMMA) was detected in urine of 354 traffic wardens [68;65] using a LC/MS/MS method, but only in about 30% of the analyzed urines it resulted quantifiable.

Table 1. Biomarkers used for biomonitoring of xylene and correspondent analytical methods

Biomarker	Treatment of sample	Analytical method	Comments
Xylene in blood	SPME	GC/MS	There is a BEI (DFE= 1.5g/L). It is correlated with the environmental exposure. It is not widely used: the matrix requires invasive sampling. There are no confounding factors.
Xylene in expired air	The phase of sampling must be conducted very carefully with tested procedures	GC/FID	There is a good correlation with the environmental data. Are not available biological limit values. The smoke is a confounding factor.
Xylene in urine	SPME	GC/MS	There's a good correlation with the environmental exposure. Are not available biological limit values. It is a good biomarker both for the single isomer and for the sum of the three. Only 1% of the absorbed dose is present as urinary xylene.
MHAs in urine	Diluition (1:10 to 1:100)	HPLC/UV	The analytical method is simple and inexpensive. Is available a biological limit value (1.5 g/g creat for the sum of the three isomers). There's an interfering peak for the determination of o-MHA. The resolution of the three isomers is difficult (if not used an added solvent e.g. THF or β-cyclodextrin in the mobile phase). For environmental data <1mg/m^3 the method is not useful for a good quantification.
	Liquid/liquid extraction or SPE and after esterification of the analytes	GC/MS	The analytical method is more laborious, more expensive but provides high accuracy, high sensitivity and lower limits of detection
Minor Metabolites (DPMA, S-BMMA) in urine	Diluition	HPLC/MS/MS	The percentage of analyte present in urine is really low compared to the absorbed dose (0.0003%), so it is needed the use of analytical method very sensitive, expensive, complex. There are no studies of correlation with other biomarkers. Are not available biological limit values.
Xylene in saliva	1mL of saliva is added to an excess of sodium sulphate in vials for headspace	HS/GC/MS	Are not available biological limit values. There are too few studies in the correlation between the levels in saliva and blood levels and/or urinary data.

Xylene in Saliva

For the analysis in saliva few methods were published, both in HS/GC/MS. The choice of this instrument is linked to the low levels of xylene that are detectable in this matrix. Some authors [21] obtained a limit of detection of 1µM with accuracy between 89-97%; others [32] showed a LOD of 0.54ng/mL with accuracy between 99-105%; regardless of the setting of these analytical methods the use in the practice of biological monitoring of salivary xylene certainly requires further investigation because only one case evaluated the correlation with blood levels and cannot be considered sufficient for the purposes of protecting the health of workers as it is not yet been demonstrated a direct correlation with exposure.

Conclusion

Over the years several biomarkers have been proposed for the implementation of the biological monitoring of workers exposed to xylene. For some (MHAs, xylene in blood) are now available BEI (Biological Exposure Index) these are the biological limit values that can be usefully applied in practice to assess any risk for workers health. For other (minor metabolites of xylene, salivary xylene, xylene in exhaled air) even if they have been set analytical methods there are objective difficulties for their use in the common practice of biological monitoring:

1. The willingness to use instrumentation highly specific and selective (e.g HPLC / MS / MS).
2. The absence of biological limit values with which to compare the data obtained, for an interpretation in terms of risk to the health
3. The need for more information about a correspondence between the absorbed dose and dose recorded in the particular matrix (for example saliva or for minor metabolites).
4. In the case of indicators whose presence consists in very low percentages with respect to the absorbed dose of xylene (minor metabolites or investigations in saliva) the possibility of use in workplaces with environmental exposures at low doses is complicated.

More studies are necessary to clarify the metabolic process of the solvent and to improve knowledge regard the distribution in the body. The availability of technologically advanced equipment provides precise quantitative levels, also extremely low.

Data by literature [69] emphasize the need for quality assurance of biological monitoring data, that should be conducted by several laboratories for quality controls.

References

[1] ATSDR (Agency for Toxic Substances and Disease Registry). *Toxicological Profile for Xylenes*. Atlanta- Georgia, USA 1993.

[2] Proctor, NH; Hughes, JP; Fischman. ML. *Chemical hazards of the workplace*. 2nd edition. 1988. Philadelphia PA. Lippincott Company.
[3] Klaassen, CD; Amdur, MO. *J. Doull. Toxicology: the basic science of poisons*. 3rd edition. 1986. MacMillan Callier Ed.
[4] HSDB (Hazardous Substances Data Bank). *Xylene*. 1994. National Library of Medicine, Washington, DC
[5] Low, LK; Meeks, JR; Mackerer, CR. *Toxicol. Ind. Health*, 5, (1989), 85-105.
[6] Savolainen, H; Pfaffli, R; Helojoki, M; Tengen, M. *Acta. Pharmacol. Toxicol*. 44, (1979), 200-207.
[7] Ransley, DL. In M. Grayson and D. Eckroth, Eds. John Wiley, *Kirk-Othmer Encyclopedia of Chemical Technology*, New York, 1984, 709-744.
[8] Langman, JM. *Pathology*. 26(3), (1994), 301-9.
[9] Watson, WP; Mutti; A. *Biomarkers*, 9, (2004), 211-242.
[10] Sedivec, V; Flek, J. *Int Arch Occup Environ Health*. 37, (1976), 205-217.
[11] Engstrom, K; Riihimaki, V; Laine; A. *Int Arch Occup Environ Health*, 54, (1984), 355-363.
[12] Wallace, L; Buckley, T; Pellizzari, E; Gordon, S. *Environ Health Persp*, 104(suppl 5), (1996), 861-869.
[13] Wallace, L; Pellizzari, ED; Hartwell, TD; Perrit, R; Ziegenfus, R. *Arch Environ Health*. 42, (1987), 272-279.
[14] Engstrom, K; Husman, K; Riihimaki, V. *Scand J Work Environ Health*, 4, (1978), 114-121.
[15] ACGIH- American Conference of Governmental Industrial Hygienists, Threshold Limit Values for chemical substances and physical agents and biological exposures indices Cincinnati, USA 2000.
[16] Elkins, HB; Pagnotto, LD; Smith, HL. *Am Ind Hyg assoc J*, 35, (1974), 559-565.
[17] Ogata, M; Taguchi, T. *Int Arch Occup Environ Health*, 58, (1986), 121-129.
[18] Ogata, M; Taguchi, T. *Int Arch Occup Environ Health*, 61, (1988), 131-140.
[19] Veulemans, H; Vlem, EV; Janssen, H; Masschelein, R. *Int Arch Occup Environ Health*, 44, (1979), 99-107.
[20] Inoue, O; Seiji, K; Watanabe, T; Nakatsuka, H; Satoh, H; Ikeda, M. *Bull Environ Contam Toxicol*. 47, (1991), 204-210.
[21] Ernstgard, L; Sjögren, B; Warholm, M; Johanson, G. *Toxicol Appl Pharmacol*, 193, (2003), 147-157.
[22] Huang, MY; Jin, C; Liu, YT; Li, BH; Qu, QS; Uchida, Y; Inoue, O; Nakatsuka, H; Watanabe, T; Ikeda, M. *Occup Environ Med*, 51, (1994), 42-46.
[23] Campbell, L; Wilson, HK; Samuel, AM; Gompertz, D. *Brit J Ind Med*. 45, (1988), 127-132.
[24] Engstrom. KM. *Scand J Work Environ Health*, 10, (1984), 75-81.
[25] Liira, J; Riihimaki, V; Engstrom, K; Pffaffli, P. *Scand J Work Environ Health*, 14, (1988), 322-327.
[26] Ikeda, M. *Toxicol Lett*, 77, (1995), 85-91.
[27] DFG- Deutsche Forschungsgemeinschaft- Commission for the investigation of health hazards of chemical compounds in the work area, list of MAK and BAT values, 2000. VCH Publichers, Weinheim, Germany.

[28] Fustinoni, S; Buratti, M; Giampiccolo, R; Colombi, A. *Toxicol Lett*, 77, (1995), 387-392.
[29] Imbriani, M; Ghittori, S; Pezzagno, G; Capodaglio, E. *Med Lav*, 78, (1987), 239-249.
[30] Caporossi, L; Santoro, A; Papaleo. B. *Biomarkers*, 15, (2010), 475-487.
[31] Erstgard, L; Sjögren, B; Warholm, M; Johanson, G. *Toxicol Appl Pharmacol*, 193, (2003), 147-157.
[32] Gherardi, M; Gordiani, A; Gatto, M. *J Chromatogr B*, 878, (2010), 2391-2396.
[33] Bergamaschi, E; Brustolin, A; De Palma, G; Manini, P; Mozzoni, P; Andreoli, R; Cavazzini, S; Mutti, A. *Toxicol Lett*, 108, (1999), 241-247.
[34] Janasik, B; Jakubowski, M; Wesołowski, W; Kucharska, M. *Int J Occup Med Environ Health*, 23, (2010), 21-6.
[35] Takeuchi, A; Kawai, T; Zhang, ZW; Miyanna, Y; Sakamoto, K; Higashikawa, K; Ikeda, M. *Int Arch Occup Environ Health*, 75, (2002), 387-393.
[36] Alkalde, TK; do Carmo Peralba, M, Alcaraz Zini, C; Bastos Caramao, E. *J of Chromatogr A*, 1027, (2004), 37-40.
[37] Gobba, F; Ghittori, S; Imbriani, M; Maestri, L; Capodaglio, E; Cavalleri, A. *Sci Total Environ*, 199, (1997), 3-12.
[38] Glaser, RA; Arnold, JE; Shulman, SA. *Am Ind Hyg Assoc J*. 51(3), (1990), 139-50.
[39] Chen, ML; Chen, SH; Guo, BR; Mao, IF. *J Environ Monit*. 4, (2002), 562-6.
[40] Angerer, J; Schaller eds, KH. o- m- p- methylhippuric acids. In: Analysis of hazardous substances in biological materials. VCH, Weinheim (FRG). 165-175.
[41] Buratti, M; Caravelli, G; Pellegrino, O; Xaiz, D; Valla, C; Giampiccolo, R; Colombi, A. *Med Lav*. 79, (1988), 489-499.
[42] Buratti, M; Pellegrino, O; Valla, C; Fustinoni, S; Brambilla, G; Colombi, A. *J Chromatogr B*, 723, (1999), 95-104.
[43] Caporossi, L; De Rosa, M; Papaleo, B. *J Chromatogr B*, 878, (2010), 2611-2615.
[44] Inoue, O; Seiji, K; Watanabe, T; Chen, Z; Huang, MY; Xu, XP; Qiao, X; Ikeda, M. *Am J Ind Med*, 25, (1994), 697-708.
[45] Kataoka, H; Manabe, K; Nakase, S; Makita, M. *J Pharm Biomed*. 9, (1991), 699-704.
[46] Sperlingova, I; Dabrowska, L; Stransky, V; Tichy, M. *Anal Bioanal Chem*, 378, (2004), 536-543.
[47] Astier, A. *J Chromatogr B*, 573, (1992), 318-322.
[48] Miller, MJ; Edwards, JW. *Int Arch Occup Environ Health*, 72, (1999), 89-97.
[49] Masana, O; Toyohiro; T. *Int Arch Occup Environ Health*, (1986), 58; 121-129.
[50] Onchoio, C; Kongtip, P; Yoosook, W; Chantanakul; S. *Southeast Asian Journal of Tropical Medicine and Public Health*, 39, (2008), 1164-1171.
[51] Buratti, M; Pellegrino, O; Caravelli, G; Xiaz, D; Valla, C; Colombi, A. *Med Lav*, 80, (1989), 254-263.
[52] Moon, DH; Paik, NW; Shim, YB. *J Chromatogr B*, 694, (1997), 367-374.
[53] Tardif, R; Brodeur, J; Plaa, GL. *J Anal Toxicol*. 13, (1989), 316-319.
[54] Fujimura, K; Ueda, T; Kitagawa, M; Takayanagi, H; Ando; T. *Anal Chem*, 58, (1986), 2668-2674.
[55] Pulley, AO; French, D. *Biochem Biophys Res Commun*, 5, (1961), 11-15.
[56] Fujii, T; Kawabe, S; Horike, T; Taguchi, T; Ogata, M. *J Chromatogr B* 730, (1999), 41-47.
[57] Lee, KJ; Lee, JJ; Moon, DC. *Electrophoresis*, 15, (1994), 95-102.

[58] Alwis, KU; Blount, BC; Britt, AS; Patel, D; Ashley, DL. *Anal Chim Acta*, 750, (2012), 152-160.

[59] De Carvalho, D; Lanchote, VL; Bonato, PS; Queiroz, CRH; Santos, AC; Dreossi; SA. *Int Arch Occup Envir Health*, 63, (1991), 33-37.

[60] Pyon, KH; Kracko, DA; Strunk, MR; Bechtold, WE; Dahl, AR; Lewis, JL. *J Anal Toxicol.* 21, (1997), 363-368.

[61] Kira, S. *Br J Ind Med*, 34, (1977), 305-309.

[62] Sollenberg, J; Baldesten, A. *J Chromatogr*, 45, (1977), 469-476.

[63] Kongtip, P; Vararussami, J; Pruktharathikul, V. *J Chromatogr B*, 751, (2001), 199-203.

[64] Van Roosmalen, PB; Drummond, I. *Br. J. Ind Med.* 35, (1978), 56-60.

[65] Sabatini, L; Barbieri, A; Indiveri, P; Mattioli, S; Violante; FS. *J Chromatogr B*, 863, (2008), 115-22.

[66] Minoia, C; Perbellini, L. "*Monitoraggio Ambientale e biologico dell'esposizione professionale a xeno biotici*" vol. 6 Solventi II. Morgan Ed. Tecniche, Como, Italy, 2001.

[67] Gonzalez-Reche, LM; Schettgen, T. *J. Angerer. Arch Toxicol.* 77, (2003), 80-5.

[68] Barbieri, A; Sabatini, L; Indiveri, P; Violante, FS. *G Ital Med Lav Ergon.* 29(3 Suppl), (2007), 536-8.

[69] Valkonen, S; Kallio, A. *J Chromatogr B*, 778, (2002), 419-427.

In: Xylenes
Editor: Michael Olawale Daramola

ISBN: 978-1-62808-342-2

Chapter 10

BIOLOGICAL TREATMENT OF XYLENE-CONTAMINATED SYSTEM: A SHORT OVERVIEW

M. O. Daramola

School of Chemical and Metallurgical Engineering, Faculty of Engineering & the Built Environment, University of the Witwatersrand, Johannesburg, South Africa

ABSTRACT

Human activities have resulted in the release of toxic chemicals into the environment and contamination of groundwater and soil worldwide. Among the major environmental contaminants, simple aromatic compounds such as benzene (B), toluene (T), ethylbenzene (EB) and xylenes are of great concern due to their relatively high water solubility and toxicity. These compounds enter the environment primarily through leakage of underground petroleum storage tanks, spills at petroleum wells, industrial effluents, wood processing, and manufacturing of pesticides, detergents, chemicals and paints. Conventionally, chemical processes have since been employed as methods for treating contaminated system (e.g. wastewater) but cost of the treatment and also generation of toxic intermediates, which in turn cause secondary pollution have been identified as major limitations. However, the use of biodegradation technology could overcome these limitations. Biological treatment technique is environmentally benign and could be cost effective when compared to the chemical treatment methods. In view of the aforementioned statements, this chapter provides a short overview of biodegradation of xylene-contaminated system. The aim is to furnish the readers with information on the existing process and also provoke thoughts of researchers towards the usage of the methods as a platform for further research development.

Keywords: Xylene, environmental pollution, biodegradation, bioremediation

INTRODUCTION

Human activities have resulted in the release of toxic chemicals into the environment and contamination of groundwater and soil worldwide. Among the major environmental contaminants, simple aromatic compounds such as benzene (B), toluene (T), ethylbenzene (EB) and xylenes are of great concern due to their relatively high water solubility and toxicity. The relatively high solubility of these compounds in water is a considerable risk for ground water contaimination. Benzene, Toluene and Xylene, commonly referred to as BTX, are natural constitutents of crude oil and gasoline. These compounds are also found in surface and ground water as a result of leakage from storage tanks or pipelines, improper disposal practices, inadvertent spills, and leaching from landfills area [1-4]. Discharges into waterways and spills on land result primarily from use, storage, and transport of petroleum products and waste disposal. However, based on the mobility of these substances in soil, xylenes may also leach into groundwater, where they may persist for several months. On the other hand, effluents from petrochemical industries are increasing significantly. The main effect that is presenred by BTX in gasoline is co-solvency, which increases the solubility of aromatic hydrocarbons in water sources. Therefore, contamination of water supplies by gasoline amended BTX and other petroleum-derived hydrocarbons is a serious and widespread environmental problem [5]. BTX compounds are dangerous carcinogens for humans, and expose environmental problems in water, soil, and indoor and outdoor air, as well as, create health risks for humans and animals [6]. When organic pollutants like benzene, Toluene, Ethylbenzene, and Xylenes (BTEX) are released into the environment, the function and structure of the microbial communities are generally affected [7,8].

Xylene has been found to bioaccumulate to very modest level (e.g. bioconcentration factor < 100), and food-chain biomagnifications has not been observed. As is expected based o the partitioning and environmental fate of xylenes, these substances are primarily detected in air. Typical xylene concentrations range from 1 to 30 ppb in outdoor air snf from 1 to 10 ppb in indoor air. According to groundwater surveys conducted in the United States, xylenes are generally detected in $< 5\%$ of groundwater samples. However, xylene concentrations in contaminated groundwater have been reported as high as 10,000 ppb. Because individual xylene isomers are used in large amounts in industrial settings, people who work at or live near these locations may receive a higher exposure to one xylene isomer compared to the other isomers. However, since xylenes are present as a mixture in gasoline and in the solvent components of commonly used commercial products (e.g. paints), exposure of the general population is expected to be primarily to xylenes as a mixture, and not to the separate xylene isomers. Xylene or its metabolites have been detected in human urine, blood, and air samples among members of the general population. Human exposure to xylene is believed to occur via inhalation of polluted indoor and workplace air, inhalation of automobile exhaust, ingestion of contaminated drinking water, smoking, and inhalation and dermal absorption of solvents containing xylene.

Conventionally, chemical processes have since been employed as methods for remediation of environmental pollution (water, air, and land). Several methods such as, advanced oxidation processes (AOP), supercritical water oxidation, UV-assisted ozonation (UV/O_3), ozone assisted photochemical oxidation ($UV/O_3/H_2O_2$), photo-fenton oxidation ($UV/H_2O_2/FeSO_4$), ozone-assisted photo-fenton oxidation ($UV/O_3/$ $H_2O_2/FeSO_4$) and radiation treatment using gamma-ray have been studied and applied [9-12]. Cost of treatment and generation of toxic intermediates and sludge, which in turn cause secondary pollution, have been identified as major limitations of these methods. However, biological technology has been proposed and studied to overcome these limitations. Biological treatment is environmental friendly and could be cost effective and energy-efficient in comparison with the chemical processes. Therefore, this chapter provides a short overview of biological process for treating xylene-polluted media. The aim is to furnish the readers with useful information about viability of this process for treatment of xylene-polluted media and to stimulate their thoughts towards the evaluation of potential application of such techniques in industries. In addition, the review might provide a platform upon which further research development in the area of biodegradation is built.

BIODEGRADATION OF XYLENE CONTAMINATED MEDIA

Biodegradation is an important process in subsurface soils and groundwater where volatilization is hindered. It is environmentally benign and could be cost effective and energy efficient in comparison with the chemical methods. In biodegradation, poor degradation of the intermediate could pose problem but the problem does not constitute a serious hindrance to its application in wastewater treatment [13-15]. It is well known that the microbial biodegradation of BTEX occurs under both aerobic and anaerobic conditions[16-19]. Advantages of aerobic treatment are high purification efficiency (>90%), high process stability and rapid biodegradability of all compounds. In recent times, anaerobic treatment has been preferred to aerobic method for the treatment of waste water due to the reasons presented in Table 1 [20].

Table 1. Comparison between aerobic and anaerobic biodegradation method. HRT: hydraulic retention time [20]

Aerobic biodegradation	Anaerobic biodegradation
High energy intensive	Low energy intensive
Large volume of sludge	Small volume of sludge
Requires highly skilled operation and process control	Requires moderate skilled operation and process control
Additional nutrients is required	No additional nutrients required
Costly technical specifications	No costly technical specifications
Poor solid settleability	Good solid settleabiblity
High HRT	Low HRT
Low elimination rate	High elimination rate

Anaerobic Biodegradation of Xylene Wastewater

Application of anaerobic degradation of BTX is not relatively new (about two decades ago) but it is constantly generating new insights. Anaerobic degradation of xylene-contaminated system has been reported under nitrate-, iron-, manganese- and sulphate-reducing conditions as well as methanogenic conditions [21]. Consequently, it has been understood that some key enzymatic steps in anaerobic biodegradation of hydrocarbon involve novel biochemistry and an in-depth knowledge of microbiota [21]. Studies of microbial degradation of monoaromatic hydrocarbons have resulted in the identification and isolation of a number of different anaerobic bacterial strains capable of degrading one or more monoaromatic hydrocarbons. Such cultures have initiated the elucidation of degradative pathways, intermediates, and genes ebcoding key enzymes. At the same time, pathways and metabolites have been inferred from complementary studies of mixed population using analytical chemical methods to demonstrate relevance of laboratory results in situ. Furthermore, evidence for anaerobic degradation of benzene, the most stable aromatic hydrocarbon, has been provided by studies with enriched populations in or from sediments containing different electron acceptors. However, mechanism of the initial reaction of benzene degradation in the absence of molecular oxygen is not yet known [22]. The occurrence and rate of biodegradation appears to be more site-specific for benzene than for other BTX components and is subject to inhibition by those co-contaminants degradation, which is usually slow, incomplete and conditioned by long lag times [21].

Anaerobic biodegradation of individual BTX components often consisted of three distinct phases [23]. Scheme describing the three phases is depicted in Figure 1. The lag phase is the phase where little or no biodegradation occurs. In the rapid degradation phase, rapid degradation of the waste is experienced. The last phase is asymptotic phase where contaminant concentrations remained essentially constant. This pattern of biodegradation cannot be predicted accurately with a simple first-order decay function, but in contrary to the behavior of the individual compounds, the biodegradation of total BTX appears to be more closely approximate by first-order decay function [23]. In most cases, Monod Equation describes biodegradation of organic compounds in sewage treatment systems, whereby the degree of compound removal depends on its concentration and activity of microbial biomass. This dependence could be described by the Equation of the first-order with respect to substrate concentration [24]. Therefore as a result of the low concentration of trace substances, the degradation is treated mostly as a first order reaction, including sorption on suspended solids yields.

Generally, some factors are considered to ensure suitability and viability of anaerobic biodegradation method for the treatment of a certain contaminated medium. Factors like bioavailability & biodegradability, microbial genome and bioreactor selection and design are very important.

Bioavailability and Biodegradability

Degradation of hazardous aromatic compounds in bioprocess depends on their bioavailability. Bioavailability decreases as transfer rate to the gas phase increases. Additionally, bioavailability may decrease as transfer to the solid phase increases, even if there is still some uncertainty about whether adsorption to bacterial surfaces hinders or aids biodegradation [25,26]. These paradigms are probably valuable for most aerobic bioprocesses including membrane bioreactor (MBR), even though the exact laws governing the relation between accessibility and biodegradation rates are still controversial. However, Barret et al. have observed that biodegradation rates decerease with increasing molecular weight in an anaerobic reactor [25]. In the case of aerobic, biodegradation rates have been found to be higher in the cross-flow MBR compared to the dead-end submerged MBR. The improvement is attributable to : (i) increase in pollutant bioavailability due to lower transfer rates to the gas and solid phases, and (ii)reduction of transfer limitation due to biomass dispersion. In some cases, degradation rate could be expected to be influenced by the active biomass concentration [25].

In a case study on anaerobic biodegradation of terephthalic acid (TA) wastewater diagnosed by Daramola et al. [20], bioavailability and biodegradability of the TA wastewater for degrading bacteria were found to be a key factor in the assessment of the viability of anaerobic biodegradation method to treat the wastewater. P-xylene is one of the constituents of TA wastewater, a waste during wet oxidation of p-xylene into acetic acid to produce crude terephthalic acid (CTA), the first step in the production of PTA. In the second step, CTA is hydrogenated to PTA of over palladium catalyst [27,28]. Besides p-Xylene, wastewater from PTA synthesis also consists of *p*-toluic acid (*p*-Tol), benzoic acid (BA), 4-carboxybenzaldehyde (4-CBA), phthalic acid (PA) and terephthalic acid (TA) with minor concentrations of 4-formylbenzoic acid, methyl acetate and *p*-xylene [29-31]. The two acid groups of in the benzene ring of TA are responsible for high solubility of TA ($\approx$ 140 $g.L^{-1}$ at pH > 5.5)[32]. In actual fact, contaminants from PTA wastewater have been associated with acute toxicity, sub-acute toxicity, chronic toxicity and molecular toxicity [33-37].

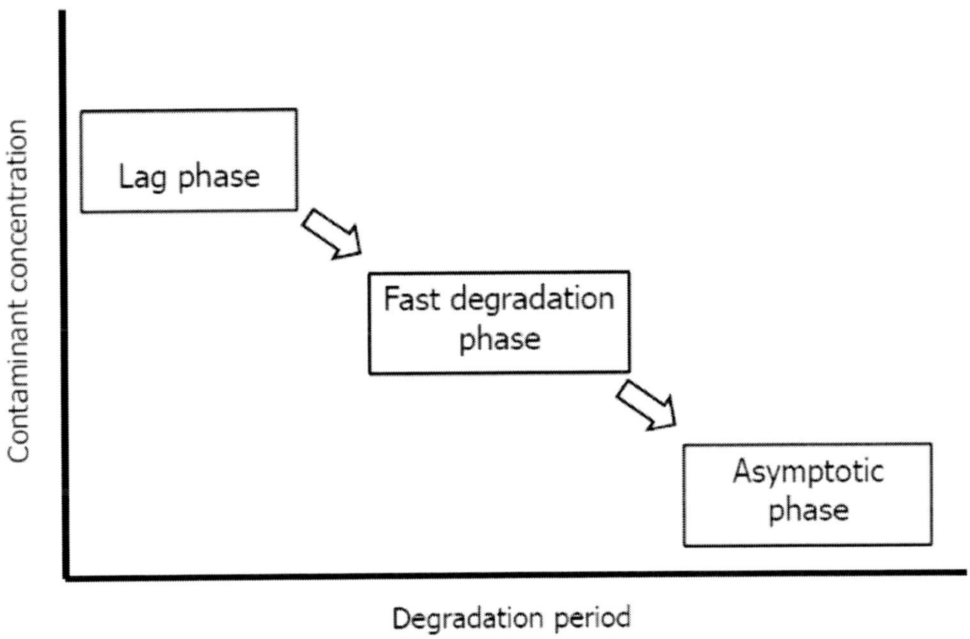

Figure 1. Phases during anaerobic biodegradation of BTX components in wastewater (adapted from ref.[23]).

Also it has been documented that biodegradability and bioavailability of a contaminant in wastewater is influenced by the pH of the waste medium [38]. Rate of growth of the degradation organisms also depends on the pH of the medium [28]. For instance, biodegradability and bioavailability of PTA wastewater (containing p-Xylene, p-toluic acid (p-Tol), benzoic acid (BA), 4-carbonxybenzaldehyde (4-CBA), phthalic acid (PA) and terephthalic acid (TA), 4-formylbenzoic acid and methyl acetate)is favoured at pH >5.5, but at pH<5.5 precipitation of the contaminants occurs, making the contaminants unavailable anymore for degradation [20]. As a result of the pH dependence of anaerobic biodegradation of TA wastewater, two-step biodegradation pathway was proposed [38]. The first step involves the decarboxylation of TA to form benzoate and the second step is the transformation of benzoate to carboxycyclohexane. With the two-step biodegradation, the growth of the TA degradation organisms was optimal st pH=6-7 [28]. At the optimal pH growth range, anaerobic biodegradation the TA wastewater was feasible. At times, adjustment of the pH of the medium is necessary to enhance biodegradability and bioavailability of a particular wastewater and addition of appropriate amount of chemicals like sodium hydroxide solution (NaOH) and sodium bicarbonate ($NaHCO_3$) to the wastewater bioreactor could push the pH of the substrate to the desired level.

Selection of Microbial Genome for Biodegradation

Identification of structure and diversity of a microbial community is important in biodegradation processes for better understanding of the physiological roles of different species in the biodegradation process. Different types of bacteria have different roles under different conditions. For instance, in thermophilic and mesophilic anaerobic biodegradation process, microbial community that is involved will exhibit different roles. For mesophilic biodegradation process, a temperature range of 35-37°C is necessary while for thermophilic anaerobic biodegradation, temperature of about 55-60°C is essential [39]. In addition, it is understood that thermophilic methanogenic consortia often display a higher specific organic removal rate than the mesophilic consortia, but to obtain thermophilic organisms in their original inoculums to provide the required conditions for thermophilic anaerobic degradation is a difficult and thus, of utmost importance [40,41]. For wastewater that contains xylenes (e.g. TA wastewater), *δ -Proteobacteria* and the *subcluster Ih* of the group 'Desulfotomaculum lineage I' have been identified as the biodegradation microorganisms under mesophilic condition [42]. Also, syntrophic methanogenic counterparts identified as *Methanosaeta concilii* and members of *Methanospirillum* and *Methanobacteriaceae* have been identified as key players in the biodegradation process.

In the case of the treatment of wastewater that contains BTEX, some natural microorganisms (either in the activated sludge or contaminated soil) have been identified to possess the ability for the biodegradation of BTEX [4]. The ability of natural microorganisms to biodegrade BTEX was first demonstrated by Gray and Thornton in 1928 and the authors identified about 250 species of bacteria present in non-contaminated environment samples applicable to environmental remediation of BTEX [43]. In addition, a variety of bacterium like Phanerochaete chrysosporium, Rhodococcus sp., Pseudoxanthomonas spadix, and Pseudomonas putida have been reported for biodegradation of BTEX [4, 44-46]. Effect of inoculation on the biodegradation performance of biodegradation orgranisms has been elucidated. For example, Lee et al. have elucidated the effect of inoculation on the biodegradation efficiency of Pseudomonas putida AY-10 during biodegradation of BTEX contaminated system [47]. The authors concluded that Pseudomonas putida AY-10 could be used as promising microbial inoculants for the direct stimulation of plant biomass production, and to degrade BTEX by plants and bacterium in the bioremediation of BTEX-contaminated environments.

Bioreactors for Biodegradation of Xylene-Contaminated System

Kinetics of biodegradation of xylene-contaminated system is essential to determining the volume and hydraulic retention time (HRT) of the bioreactor required for biodegradation process. In order to study the kinetics of biodegradation, in many cases, some assumptions are made due to insufficient data from literature. These assumptions are justified and validated through experimentations to obtain the kinetic parameters for design purpose. Additionally, in-depth investigation of anaerobic biodegradation mechanism and bio-kinetics is also essential.

Several bioreactors have been developed and studied for anaerobic biodegradation of contaminated system. In the 1970's, Lettinga and his co-workers at Wageningen University in The Netherlands developed Upflow anaerobic sludge blanket (UASB) reactor for anaerobic degradation of wastewater (see Figure 2). Also, some research efforts also have been carried out with the use of biofilm reactors with attached growth medium for microbes to degrade wastewater [38,41,48]. Furthermore, in the laboratory studies of biodegradation of contaminated system, different experimental set-ups have been developed and used for experimental biodegradation. Down flow tubular fixed film reactor has been proposed and used by Noyola et al. [49] for PTA wastewater treatment (see Figure 3). The set-up was made with plexiglass column of 1 m high and 96 mm internal diameter and packed with 21 polyvinyl chloride (PVC) tubes of each, 670 mm high and 12.7 mm diameter. In total, the reactor could provide 1.05 m^2 support area and a void volume of 4.75 [49]. In comparison to UASB reactors, Downflow tubular fixed-film reactor has good resistance to shock loads and periods without feeding. The aforementioned factors could be major limitations in anaerobic biodegradation of xylene-contaminated wastewater.

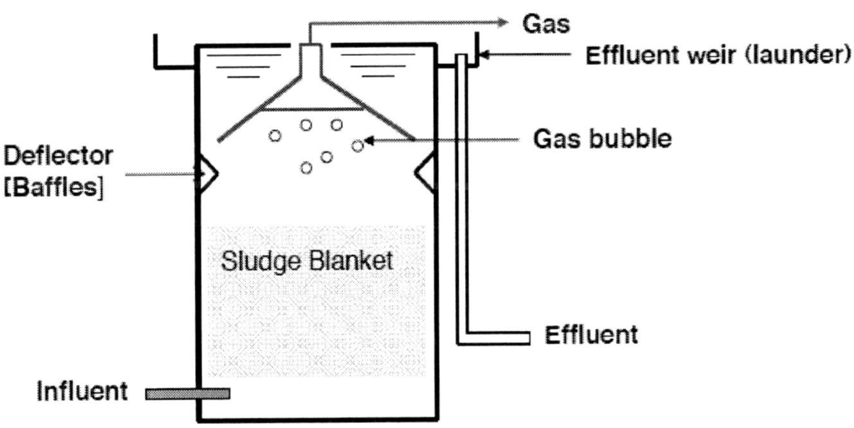

Figure 2. Upflow anaerobic sludge blanket (UASB) reactor.

In the same vein, Pophali et al. [50] have investigated anaerobic biodegradation of wastewater using a laboratory scale Upflow Anaerobic Fixed-film Fixed-bed Reactor

(AFFFBR) (see Figure 4). The reactor was made of glass with reactor volume 2.43 L and about 1.45 L of the total volume of the reactor was occupied with 10-20 mm (size) stones resulting in void volume 0.98. Using the Upflow Anaerobic Fixed-film Fixed-bed Reactor (AFFFBR) to degrade xylene-contaminated wastewater, Saghafi et *al.* demonstrated that removal of chemical oxygen demand (COD) decreases as HRT decreases [51]. For instance, at HRT of 24 h and COD concentration of 3000 mg/l, the COD removal efficiency of the bioreactor was 62 % but the efficiency decreased to 42 % and 30% when the HRT was reduced to 16 h and 8 h, respectively (see Figure 5).

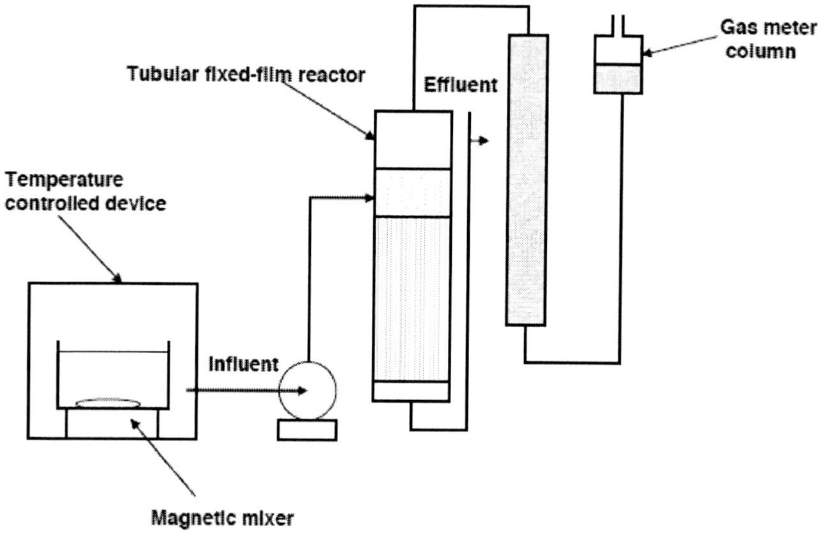

Figure 3. Downflow tubular fixed-film reactor (Adapted from ref. [49]).

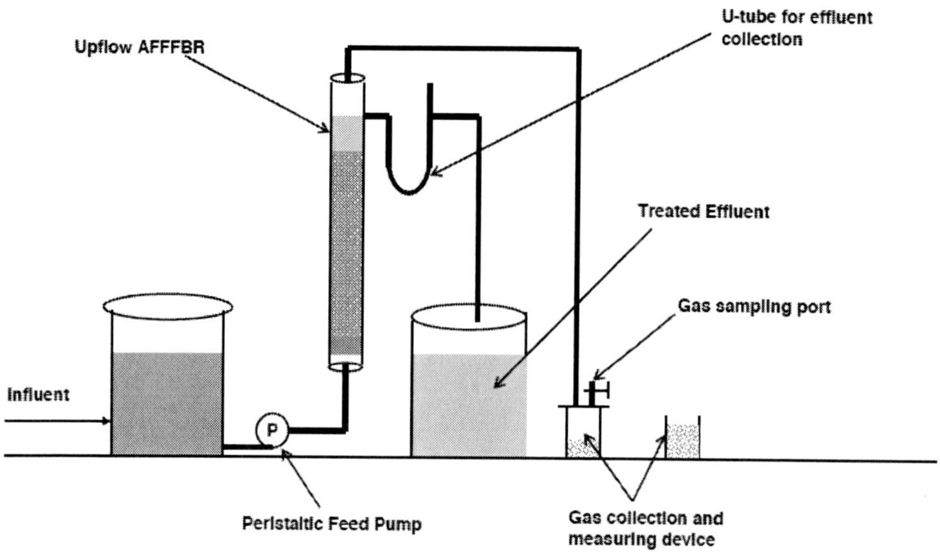

Figure 4. Upflow anaerobic-aerobic fixed-film fixed-bed reactor (UAFFFBR) (Adapted from ref. [50]).

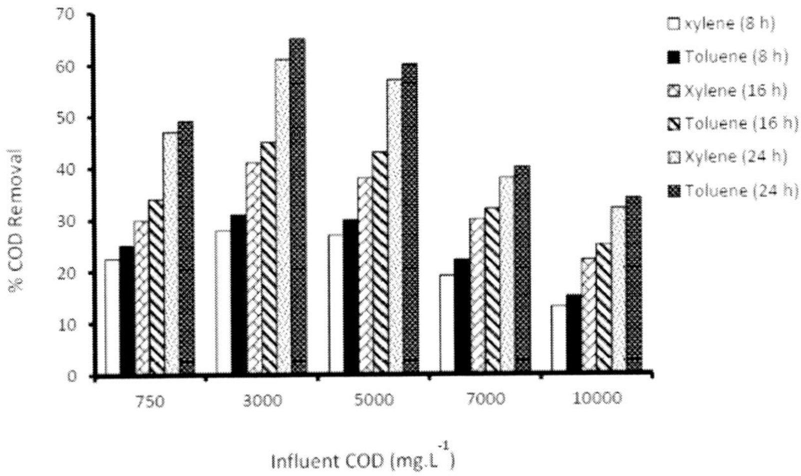

Figure 5. Influence of HRT on the COD removal efficiency during biodegaradtion of xylene polluted wastewater in UAPB reactor (adapted from ref. [51]).

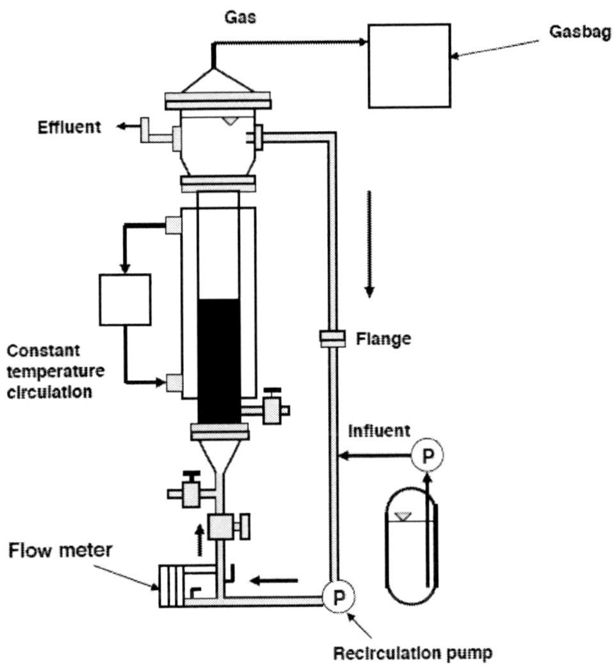

Figure 6. An expanded-bed granular activated carbon (GAC) anaerobic reactors (Adapted from ref. [52]).

Factors limiting the efficiency of anaerobic degradation of wastewater are chemical inhibition effects and shock loads [49]. To alleviate these problems, expanded-bed granular activated carbon (GAC) anaerobic reactors have been developed and implemented to investigate the treatment of xylene-contaminated waste water [52-55]. Figure 6 depicts a typical example of an expanded-bed granular activated carbon (GAC) anaerobic reactor. The reactor is made of a thick Plexiglas tube with a diameter of 100 mm, occupying 10 L volume.

1.5 kg GAC with a particle size of 0.9–1.1 mm is placed in the reactor as the attached growth medium. The liquid in the reactor is circulated from the top part to the bottom to expand the GAC medium by 25%, resulting in an expanded-bed volume of 4.3 L. The concept behind the mechanism of the reactor is based on the physical and biological removal mechanisms. In the physical removal mechanism, adsorption of inhibitory chemicals is by adsorption onto GAC while the biological removal is through degradation by microorganisms growing on the GAC [52]. Through the adsorptive function of the GAC, effect of the inhibitory chemicals and effect of shock loads are minimized [52]. Other bioreactors that have been studied for anaerobic biodegradation of xylene-contaminated wastewater include anaerobic sequencing biofilm batch reactor (ASBBR) [56] and Horizontal-flow anaerobic immobilized biomass reactors (HAIB) [57]. Despite series of reactors that have been proposed and developed, there is still need for further research efforts toward optimizing and up-scaling of these reactors for anaerobic treatment of xylene-contaminated wastewater.

Biological Treatment of Xylene-Contaminated Air

Air pollution is the introduction into the atmosphere of chemicals, particules, or biological materials that cause discomfort, diseases, or death to humans, damage other living organisms, such as food crops, or damage the natural environment or built environment. Also, the atmosphere is a complex dynamic natural gasesous system that is essential to support life on the planet Earth [58] and pollutants are substances in the air that can be harmful to humans and the environment. These substances, which are either natural or man-made, can be in form of solid particles, liquid droplets, or gases. Pollutants are classified into primary and secondary forms. Primary pollutant are directly emitted from a process, such as ash from volcanic eruption, the carbon monoxide gas from a motor vehicle exhaust or sulphur dioxide released from factories. Secondary pollutants are not emitted directly, but they form in air when primary pollutants react or interact (e.g ground ozone, one of the many secondary pollutants that make up photochemical smog). According to the World Health Organization (WHO), about 2.4 million people die every year from causes related to air pollution [59].

Xylenes are released to the atmosphere primarily as fugitive emissions from industrial sources (e.g. petroleum refineries, chemical plants), in automobile exhaust, and through volatilization from their use as solvents. Xylenes released to soil or surface water are expected to volatilize into the atmosphere and rapidly transformed in the troposphere where photo-oxidation by hydroxyl radicals occurs. Therefore reduction/removal of xylene from contaminated air is essential to reduce the health risk to the occupants of the Earth.

Due to the environmental benign of the biological treatment of pollution compared to the chemical methods, the use of biofilters has been adopted and developed for the removal of xylene from air and optimization studies of such devices is necessary. Wu et al. have demonstrated the efficiency of an hybrid biofilter for the removal of Xylenes from air [60]. The schematic of the hybrid biofilter is presented in Figure 7. The hybrid biofilter system consists of two biofilters including a compost-based A and a polyethylene (PE)-based biofilter B. The compost-based biofilter (A) is packed with mixture of mature pig compost, forest soil and polyethylene-based material and the PE-based biofilter B is also packed with the polyethylene-based material. Each biofilter was made from a height of 0.33 m cylindrical polymethyl methacrylate column with inner diameter of 0.05 m, and filled to a height of 0.17 m. More details of the hybrid biofilter can be obtained from ref. [60]. The authors reported about 100% removal efficiency for the system, even at increasing concentration of the p-Xylene. The overall performance of the hybrid biofilter during the treatment of xylene-contaminated air is depicted in Figure 8.

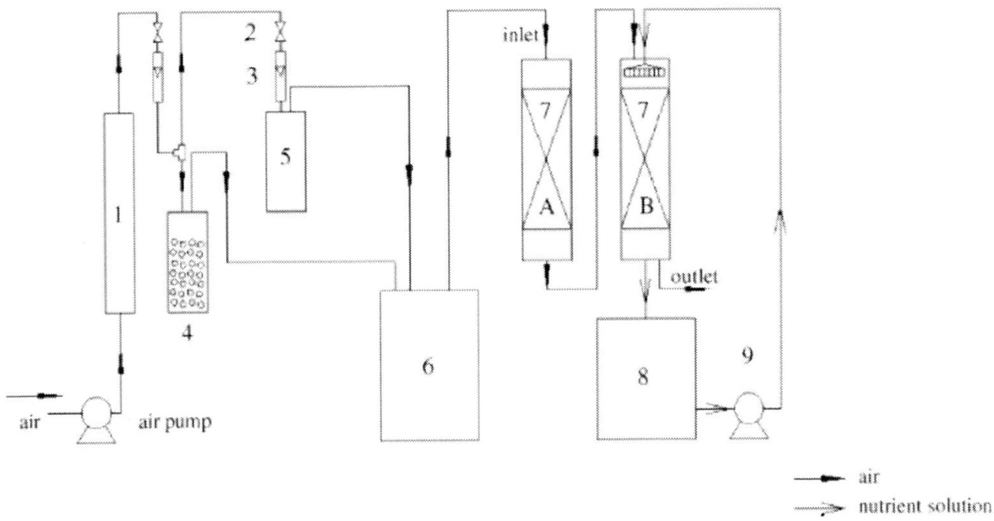

Figure 7. Schematic of hybrid biofilter system (obtained with permission from ref. [60]).

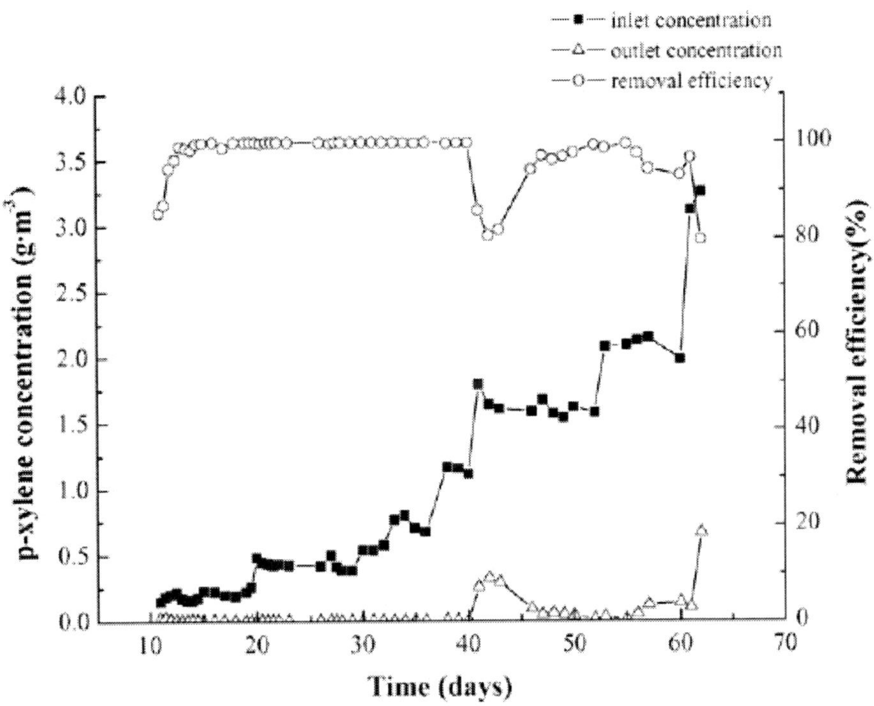

Figure 8. Overall performance of hybrid biofilter during the treatment of xylene-contaminated air (obtained with permission from ref. [60]).

Bioremediation of Xylene-Contaminated Soils

Bioremediation is the process of cleaning up hazardous wastes with the use of microorganisms or plants, and it is the safest method of cleaning soil of pollutants [61]. Bioremediation of soil contaminated with toxic chemicals is the removal of, and/or reduction in, the amount of toxic substances to ensure the soil is not harmful to animals, humans, plants and beneficial soil microbes. In soil, toxic aromatic hydrocarbons may sometimes exist in an adsorbed state, but nevertheless may still participate in chemical reactions and can be removed by bioremediation. Heath consequences from exposure to soil contamination vary greatly depending on pollutant type, pathway of attack and vulnerability of the exposed population. Chronic exposure to Chromium (Cr), Lead (Pb) and other metals, petroleum, solvents, and many pesticide and herbicide formulations can be carcinogenic and can cause other chronic health conditions [62].

During bioremediation, microbes utilize chemical contaminants in the soil as an energy source and, through oxidation-reduction reactions, metabolize the target contaminant into usable energy for microbes. After bioremediation, the metabolites or by-products released into the environment are less harmful and toxic to the environment. Bioremediation depends on (i) presence of a contaminant, (ii) an electron acceptor, and (iii) presence of microorganism that are capable of degrading the specific contaminant [63]. Generally, some added nutrients, such as nitrate, sulphate, ferric iron, manganese, or carbon dioxide serve as terminal electron acceptors to facilitate biodegradation [63].

Research efforts have shown that Gram-negative bacteria of the family *Rhizobiaceae*, including *S. Meliloti* (formerly Rhizobium meliloti) are useful for bioremediation of xylene-contaminated soil, such as reduction of 2,4,6-trinitrotoluene (TNT) into amino-Dinitrotoluene (DNT)[64,65]. Other products derived from TNT are DNT, Toluene (T), benzene (B), ethylbenzene (EB) and xylenes. These toxic aromatics are manufactured for use in fertilizers, herbicides, and pesticides, as well as in the production of plastics and dyes and they contaminate soils in the environment as a result of agricultural run-off, industrial wastes sites, and military operations [66,67]. Additionally, *S. Meliloti* is a motile, rod-shaped bacterium which naturally lives as a microbial constituent of soil microbial populations are more amenable to genetic manipulation than many other legume-nodulating bacterial species [64]. A recent study has demonstrated the enhanced efficiency of genetically-engineered *S. Meliloti* (S. Melilotis strain USDA 1936) in rhizosphere bioremediation of TNT [68]. The authors genetically modified S. meliloti through intergeneric transfer of plasmid pJS1 to *S. meliloti* strain USDA 1936 from *Burkholderia sp. Strain DNT* and evaluated the intergeneric transconjugant in the biodegradation of 2,4-DNT via alfalfa/S. meliloti symbiosis. Alfalfa is a leguminous plant with extensive soil-root systems extending as deep as 2 m, enabling the plant to act both as a vehicle for limiting the spread of solutes and promoting the mineralization of the toxic compound [69]. Furthermore, the genetically modified *S. meliloti*, pJS1 plasmid, carries all the necessary genes for 2,4-DNT degradation [68]. The results from the studies also showed that the pJS1-carrying recombinant strain DHK1 produced "ARC" alfalfa plants that were almost 2-fold higher in shoot dry weight than that produced by the parent strain on soil containing 0.14 mM 2,4-DNT. Furthermore, the trans-conjugant strain DHK1 degraded about 94% of the 2,4-DNT present in 0.55 mM contaminated soil.

Factors affecting rate of microbial bioremediation of a contaminated soil have been identified as (i) availability of nutrient, (ii) moisture content, (iii) pH of the contaminated medium, and (iii) temperature of the soil matrix [70]. Inorganic nutrients, such as nitrogen, phosphorus are necessary for microbial activity and cell growth. Also all soil microorganisms require moisture for cell growth and function. Presence of water during soil bioremediation affects diffusion of water and soluble nutrients into and out of the microorganism cells. Also, excess moisture (in case of saturated soil) reduces amount of available oxygen for aerobic bioremediation and in this case, anaerobic bioremediation dominates. Soil pH is very important because most degrading microbes during bioremediation can survive within a certain pH range and pH does affect availability of nutrients for bioremediation. For example, acceptable pH range during bioremediation of petroleum hydrocarbon contaminated soil is 6 to 8 [71]. Rate of biodegradation is influenced by temperature because temperature controls the rate of enzymatic reactions within the microorganisms. It has been reported that rate of enzymatic reactions in a cell doubles (approximately) for every 10°C rise in temperature [63,72]. Furthermore, soil type should be considered when determining the best suited bioremediation approach to a particular situation. In this regards, bioremediation can be categorized into *in situ* bioremediation referring to treatment of contaminated soil in pace, and *ex situ* bioremediation, which involves the use of bioreactors, landfarming and biophiles to treat contaminated soil [72]. *In situ* bioremediation causes minimal disturbance to the environment and is cost effective compared to conventional bioremediation technique because no transportation or replacement of soil is required. However, *in situ* bioremediation is not suitable for all soils, leads to incomplete degradation and difficult to control and optimize. Meanwhile in both methods, to achieve optimal bioremediation constant monitoring and regular adjustment of operating conditions are essential [74].

In conclusion, bioremediation offers an environmentally friendly and cost competitive approach for the clean-up of xylene-contaminated sites, but the technology is slower than other technologies [75]. Also, bioavailablity of pollutants and degradation competence of the indigenous microbiota might be limiting factors in treating aged historically polluted soils [76].

CONCLUSION

In this chapter, various biological techniques for the treatment of xylene-contaminated environment/system, such as xylene-contaminated water, air and soils, have been enumerated. It is essential to obtain reliable monitoring data for levels of xylene in a contaminated medium at hazardous waste sites. The acquired information on levels of xylene in the environment can be used in combination with the known body burden of xylene to assess the potential risk of adverse health effects in populations living in the vicinity of the hazardous waste sites. Additionally, in-depth understanding of effects of influencing factors, such as microbial genome, type of bioreactors, thermodynamics, bioaugmentation and other additives likes non-ionic surfactants on the degrading capability of autochthonous microbiota (in case of soil bioremediation) should be sought to encourage large-scale development of the biological techniques for the treatment of xylene-contaminated system.

Acknowledgment

The author acknowledges Elsevier publishing BV, The Netherlands for the permission to use some figures from its published work in this chapter.

References

[1] Heider, J; Spormann, AM; Beller, HR; Widdel, F. *FEMS Microbiol. Rev*, 22, (1999) 459-473.
[2] Lovley, DR. *Biodegradation*, 11, (2000), 107-116.
[3] Phelps, CD; Young, LY. *AdvAgron*, 70, (2001), 329-357.
[4] Mazzeo, DE; Levy, CE; deF, D; de Angelis, MA. *Marin-Morales Sci. Total Environ.* 408, (2010), 4334-4340.
[5] Gusmao, VR; Martins, TH; Chinalia, FA; Sakamoto, IK; HenriqueThiemann, O. M.B.A Varesche, *Process Biochem.* 41, (2006), 1391-1400.
[6] Corseuil, HX; Hunt, CS; Ferreira dos Santos, RC; Alvarez, PJ. *Wat. Res.* 32, (1998), 2065-2072.
[7] Barker, JF; Patrick, GC; Major, D. *Ground Water Monit. Rev.* 7, (1987), 64-71.
[8] Lovely, DR. *Science*, 293, (2001), 1444-1446.
[9] Deng, Y; Zhang, K; Chen, H; Wu, T; Krzyaniak, M; Wellons, A; Bolla, D; Douglas, K; Zuo, Y. *Atmos. Environ.*, 40, (2006), 3665–3676.
[10] Chan, AH; Chan, CK; Barford, JP; Porter, JF. *Water Res.*, 37, (2003), 1125–1135.
[11] Brillas, E; Cabot, PL; Rodriguez, RM; Arias, C; Garrid, JA; Oliver, R. *Appl. Catal.*, 51, (2004), 117–127.
[12] Radoiu, MT; Martin, DI; Calinescu, I; Iovu, H. *J. Hazard. Mater.*, 106, (1), (2004), 27–37.
[13] Kleerbezem, R; Beckers, J; Hulshoff Pol, LW; Lettinga, G. *Biotechnol. Bioeng.*, 91(2), (2005), 169–179.
[14] Lau, CM. Staging aeration for high efficiency treatment of aromatic acids plant wastewater, Proceedings of 32nd Independent Wastewater Conference, Purdue University, West Lafayette, Ind., 1978, 63-74.
[15] Macarie, H; Guyot, JP. *Environ. Technol.*, 16, (1995), 1185–1192.
[16] Edwards, EA; Wills, LE; Reinhard, M; Grbic-Galic, D. *Appl. Environ. Microbiol.* 58, (1992), 794-800.
[17] Evans, PJ; Mang, DT; Young, LY. *Appl. Environ. Microbiol.* 57, (1991), 450-454
[18] Hutchins, SR. *Appl. Environ. Microbiol.* 57, (1991), 2403-2407.
[19] Smith, MR. *Biodegradation*, 1, (1990), 191-200.
[20] Daramola, MO; Aransiola, EF; Adeogun, AG. *Nat. Sci.* 3(5), (2011), 371-378.
[21] Foght, J. *J. Molecular Microbiol. & Biotech.* 15, (2008), 93-120.
[22] Widdel, F; Rabus, R. Anaerobic biodegradation of saturated and aromatic hydrocarbons, *Current option in Biotechnology*, 12, (2001), 259-276.
[23] Borden, RC; Hunt, MJ; Shafer, MB; Barlaz, MA. *Anaerobic biodegradation of BTEX in aquifer material*, United States Environmental Protection Agency, EPA/600/S-97/003, 1997.

[24] Rittman, B. *The Science of the total Environment*, 47, (1985), 99-113.

[25] Barret, M; Carrere, H; Delgadillo, L; Patureau, D. *Water Research*, 44(13), (2010), 3797-3806.

[26] Stringfellow, WT; Alvarez-Cohen, L. Evaluating the relationship between the sorption of PAHs to bacterial biomass and biodegradation, *Water Research*, 33(11), (1999), 2535-2544.

[27] Franck, HG; Stadelhofer, JW. P-Xylene and its derivates: terephthalic acid, in Franck, H.G., (Ed.), *Industrial Aromatic Chemistry: Raw Materials, Process Products*, Springer-Verlag, Berlin, Heidelberg, Germany,1998, 283–290.

[28] Kleerbezem, R; Mortier, J; Hulshoff Pol, LW; Lettinga, G. *Water Sci. Technol.*, 36, (2–3), (1997), 237–248.

[29] Liangming, X; Yuxiang, C; Xiangdong, Z. In: H. Xianglin, Ed., The Anaerobic Biological Treatment of High Strength Petrochemical Wastewater by a Hybrid Reactor, *Int. Acad. Pub.*, Beijing, China, 1991, 120–126.

[30] Kleerebezem, R; Hulshoff Pol, LW; Lettinga, G. *Biodegradation*, 10(1), (1999), 63–73.

[31] Kleerebezem, R; Hulshoff Pol, LW; Lettinga, G. *Appl. Environ. Microbiol.*, 65(3), (1999), 1152–1160.

[32] Macarie, H; Noyola, A; Guyot, JP. *Water Sci. Technol.*, 25(7), (1992), 223–235.

[33] Qi, ST; Wang, XR; Xu, XK. *J. Hyp. Res.* 31(1), (2002), 10-12.

[34] Yuan, YX; Qu, KM. *J. Fish. Sci. China*, 7(2), (2000), 82-85.

[35] Qu, KM; Yuan, YX; Chen, MS. *J. Fish Sci. China*, 7(11), (2000), 78-81.

[36] Chen, BJ; Yuan, YX; Wang, HP. *J. Fish. Sci. China*, 8(1), (2001), 73-76.

[37] Zang, HD; Xu, XK; Gong, N. *China J. Ind. Med.* 14(2), (2001), 65-67.

[38] Kleerebezem, R; Pol, LWH; Lettinga, G. *FEMS Microbiol. Ecol.*, 29, (1999), 273-282.

[39] Macarie, H. *Water Sci. Technol.* 42, (2000), 201-214.

[40] Van Lier, JB; Rebac, S; Lettinga, G. *Water Sci. Technol.*, 35(10), (1997), 199-206.

[41] Chen, C-L; Macarie, H; Ramirez, I; Olmos, A; Ong, SL; Monroy, O; Liu, WT. *Microbiology*, 150, (2004), 3429-3440.

[42] Qiu, YL; Sekiguchi, Y; Imachi, H; Kamagata, Y; Tseng, IC; Ohashi, SS; Harada, H. *Appl. Environ. Microbiol.*, 70, (2004), 1617-1626.

[43] Gray, PHH; Thornton, HG; Zentralbl. *Bakteriol. Parasitenk. Infektionskr. Hyg. Abt. II.* 73, (1928), 74-96.

[44] Yadav, JS; Reddy, CA. *Appl. Environ. Microbiol.* 59, (1993), 756-762.

[45] Lee, EH; Cho, KS. *J. Hazard. Mater.* 167, (2009), 669-674.

[46] Kim, JM; Le, NT; Chung, BS; Park, JH; Bae, J; Madsen, EL; Jeon, CO. *Appl. Environ. Microbiol.* 74, (2008), 7313-7320.

[47] Lee, EY; Hong, SH; Oh, MH; Lim, JS. Characterization biodegradation of benzene, Toluene, Etheylbenzene, and Xylenes by newly isolated Bacterium Pseudomonas Putida AY-10 in Rhizosphere of wastewater Treatment Reed, 3rd International Conference on Chemical, *Biological and Environmental Engineering IPCBEE*, vol. 20, ACSIT Press, 2011, Singapore.

[48] Kleerbezem, R; Beckers, J; Hulshoff Pol, LW; Lettinga, G. *Biotechnol. Bioeng.*, 91(2), (2005), 169–179.

[49] Noyola, A; Macarie, H; Guyot, JP. *Environmental Technol.*, 11, (1990), 239-248.

[50] Pophali, GR; Khan, R; Dhodapkar, RS; Nandy, T; Devotta, S. *J. Environmental Management*, 85, (2007), 1024-1033.

[51] Saghafi, S; Bakhashi, Z; Najafpour, GD; Kariminezhad, E; Rad, HA. *American-Eurasian J. Agric. & Environ. Sci.* 9, (2010), 1-7.
[52] Tsuno, H; Kawamura, M. *WATER RESEARCH,* 43, (2009), 417-422.
[53] Suidan, MT; Strubler, CE; Kao, SW; Pfeffer, JT. *J. Water Pollution Control Federation,* 55(10), (1983), 1263-1270.
[54] Tsuno, H; Kawamura, M; Somiya, I. *Water Science and Technology*, 34(5–6), (1996), 335–344.
[55] Tsuno, H; Kawamura, M; Oya, T. *Water Science and Technology,* 53(11), (2006), 251–260.
[56] Estebar, M; Amin, MM; Poursafa, P; Ghasemian, M; Jaafarzadeh, N; Hashemi, H; Fatehizadeh, A. Biodegradation of benzene-toluene-xylene in petrochemical industries wastewater through anaerobic sequencing biofilm batch reactor in bench scale, *International J. Environ. Health Eng.* 1(2), (2012).
[57] Damianovic, MHRZ; Moraes, EM; Zaiat, M; Foresti, E. *Bioresource Technol.* 100, (2009), 4361-4367.
[58] http://en.wikipedia.org/wiki/air_pollution(accessed 15-03-2013).
[59] Estimated death & DALYs attributable to selected environmental risk factors, by WHO member states, 2002, (accessed 15-03-2013).
[60] Wu, D; Quan, X; Zhao, Y; Chen, S. *J. Hazardous Mater.* 136, (2006), 288-295.
[61] Alleman, BC; Leeson, A. In situ and on-site bioremediation, Battlelle Press, Columbus, OH, 1997.
[62] Snyder, C. *Int. J. Occup Environ Health,* 11(4), (2005), 415-427.
[63] Nester, EW; Anderson, GD; Roberts, CE; Pearsall, NN; Nester, MT. Microbiology: A Human perspective, 3rd ed., McGraw-Hill, New York, 2001.
[64] Kuykendall, LD; Hashem, FM; Buchan, GR; Devine, TE; Dadson, RB. *Symbiosis*, 27, (1999), 1-16.
[65] Labidi, M; Ahmad, D; Halasz, A; Hawari, J. *Canadian J. Microbiol.* 47, (2001), 559-566.
[66] Spanggord, RJ; Spain, JC; Nishino, SF; Mortelmans, KE. *Appl. & Environ. Microbiol.* 57, (1991), 3200-3205.
[67] Duque, E; Haidour, A; Godoy, F; Ramos, JL. *J. Bacterial.* 175, (1993), 2278-2283.
[68] Duttaa, SK; Hollowella, GP; Hashemb, FM; Kuykendallc, LD. *Soil Biology & Biochemistry*, 35, (2003), 667-675.
[69] Anderson, TA; Coats, JR. Bioremediation through Rhizosphere Technology, *American Chemical Society*, Washington DC, 1994.
[70] Walworth, J; Pond, A; Snape, I; Rayner, J; Ferguson, S; Harvey, P. *ARCSACC*, (2005), 251-257.
[71] Landfarming, US EPA, 2006.
[72] Ameda, P; Banat, IM; Marchant, R; Vezzulli, L. *Chemosphere*, 66, (2007), 179-184.
[73] Fundamental of Bioremediation, Deaprtment of Environmental Quality, State of Mississipi, USA, April 1998 & Nov. 2006, (www.deq.state.ms.us).
[74] A citizen's guide to bioremediation, Environmental Protection Agency, United States, Nov. 2006, (www.epa.gov/tio/downlaod/citizens/bioremediation.pdf.
[75] Alexander, M. *Biodegradation and bioremediation*, Academic Press, London, 1999.
[76] Boopathy, R. *J. Harzadous Mater.* 92, (2002), 103-114.

INDEX

A

B

D

Q

R

S

T

U

V

W

X

Y

Z